Innovation, Networks and Learning Regions?

Regional Policy and Development Series

Series Editor: Ron Martin, Department of Geography, University of Cambridge

Throughout the industrialised world, widespread economic restructuring, rapid technological change, the reconfiguration of State intervention, and increasing globalisation are giving greater prominence to the nature and performance of individual regional and local economies within nations. The old patterns and processes of regional development that characterised the post-war period are being fundamentally redrawn, creating new problems of uneven development and new theoretical and policy challenges. Whatever interpretation of this contemporary transformation is adopted, regions and localities are back on the academic and political agenda. *Regional Policy and Development* is an international series which aims to provide authoritative analyses of this new regional political economy. It seeks to combine fresh theoretical insights with detailed empirical enquiry and constructive policy debate to produce a comprehensive set of conceptual, practical and topical studies in this field. The series is not intended as a collection of synthetic reviews, but rather as original contributions to understanding the processes, problems and policies of regional and local economic development in today's changing world.

Innovation, Networks and Learning Regions?

Edited by James Simmie

Regional Policy and Development Series 18

Routledge
Taylor & Francis Group
LONDON AND NEW YORK

This book is dedicated to my mother, MFS, whose dining room table still bears the scars of lengthy earlier studies

First published in the United Kingdom in 1997 by
Jessica Kingsley Publishers Ltd
with
The Regional Studies Association
Registered Charity 252269

Published 2004
By Spon Press

Published 2017 by Routledge
2 Park Square, Milton Park, Abingdon, Oxon OX14 4RN
711 Third Avenue, New York, NY 10017

First issued in hardback 2017

Routledge is an imprint of the Taylor & Francis Group, an informa business

Library of Congress Cataloging in Publication Data
A CIP catalogue record for this book is available from the Library of Congress

British Library Cataloguing in Publication Data
Innovation, networks and learning regions?. - (Regional policy and development ; 18)
1. Economic geography 2. Economic policy 3. Economic development
I. Simmie, J.M. (James, Martin), 1941-
338.9

ISBN 13: 978-1-138-41088-6 (hbk)
ISBN 13: 978-0-11-702360-4 (pbk)

Contents

Part IV. Technology Transfer

Part V. Conclusions

PART I

Introduction

CHAPTER 1

Origins, Structure and Contents

James Simmie

Introduction

This book was inspired by a series of sessions on 'Innovation and Regional Development' conducted at the Regional Studies Association conference on 'Regional Futures: Past and Present, East and West' held in Gothenberg in May 1995. Innovation and economic development were themes which reappeared constantly not only in this but also in other strands of this excellent international meeting. Within these major themes some key issues emerged which were frequently addressed in subsequent discussions. This volume contains selected research which addresses these key issues from different theoretical, empirical and national perspectives.

These key issues, which form the linked focus of the substantive parts of this book, are:

- Core metropolitan regions – This part examines the role played by core metropolitan regions as the major locations and sources of innovations and their subsequent profitable development.
- Peripheral regions – This part analyses the problems faced by more peripheral regions in overcoming their comparative disadvantages with respect to innovative capacities and the public policies developed to reduce their disadvantage.
- Technology transfer – Here the focus is on the question of whether essential technology transfers can best take place in terms of information movement using electronic techniques or knowledge transfer requiring the movement of people.

The concluding part draws out the linking themes presented in the separately authored chapters. Many of these themes are concerned with the need for regions to develop and remain competitive in the global market-place by constant innovation. In seeking to achieve this goal they need to face in two directions at once. On the one hand they must build and maintain their own distinctive regional innovation systems capable of generating continuous change and on the other they must compete in global markets full of regions and firms all attempting to do much the same thing. This essential double focus is described as the 'global/local interface'.

Part II of the book is concerned with the innovative performance of firms in the south-east of England and the Greater Paris region. In Chapter 2, James Simmie reports on some preliminary results of a study of award-winning firms in Hertfordshire. This is an area, within the London western arc, which has been among the relatively most innovative in the UK. The characteristics of award-winning product and process innovations are examined – together with the local factor conditions, customer and demand conditions, related and supporting industries, and firm strategy, structure and rivalry that contributed to them.

The findings of this study suggest that local networking is not normally significant in such regions. Instead, the demand for local high quality staff is crucial. This is often combined with the ability to access international markets relatively easily from the south-east. The study also suggests that government regulations are not significant with respect to particular innovations. Industrial sectors tend to remain distinct from one another. They do, however, often rely on generalised local office support services which do not usually involve much in the way of high technology.

In Chapter 3, Pooran Wynarczyk and Alfred Thwaites contribute to the debate surrounding regional economic development and, in particular, the role of small and medium sized enterprises (SMEs) in this process. They examine empirically the influence of a number of entrepreneurial and regional factors on the financial performance of a sample of 170 innovative small firms. The empirical analysis, based upon Companies House data, examines the relative explanatory power of a number of entrepreneurial, regional and financial variables on the performance of innovative small firms in different operational environments. The sample is divided into two groups. One is composed of 71 firms located in the south-east local economic planning region and the other is made up of 99 firms located elsewhere in the United Kingdom.

The results confirm earlier findings that substantial innovations introduced by SMEs over the period 1975–1983 were concentrated in a few sectors of industry. Surviving and innovative SMEs are part of a set of fast-growing firms which warrant the attention they receive from policy makers and academics alike.

At the regional level the evidence suggests that significant innovations are more likely to be introduced into the south-east region than elsewhere in the country. The work also shows that, post-innovation, retained profits and exports grow more strongly in firms located in the south-east than those located in other regions of the UK. Whilst the majority of firms in the other regions were family-run, the south-east firms were, in contrast, run by more professional directors and were more often associated with exports and profitability growth than family-run firms.

In Chapter 4, Jeanine Cohen examines the restructuring and relocation of firms with high levels of research and development (R&D) in the Greater Paris region. She shows that, from the 1980s, the places with the highest relative levels of R&D employment were located in the west contiguous 'communes' of Paris and, more widely, in the peri-urban, south-western and southern areas. Nevertheless, the industries and firms that make up the 'high-tech' phenomenon are changing. This is due, at least in part, to the international restructuring of firms. This restructuring is also affecting the traditional industrial zones such as the northern suburbs of the capital.

Modernisation, involving a certain amount of decentralisation, is also associated with acquisitions and job cuts. The aim of this piece of research is to discover whether local milieux, such as *La Cité Scientifique* and other local actors, can counter-balance the trends driven by acquisitions, deregulation and privatisations in such a way as to modify the role of R&D in the Paris region.

In Chapter 5 Heidi Wiig and Michelle Wood report a theoretical and empirical study of a non-metropolitan region as an economic space and innovation system. They present data from a comprehensive survey of innovations among manufacturing firms in the region of Møre and Romsdal. This is a coastal region in central Norway with a wide mix of mostly small-scale manufacturing activities.

They argue that although there has been a great deal of theoretical and empirical work attempting to explain regional divergences in technological performance and economic growth, there appears to be no systematic approach to the study of innovation systems in a regional context and closer investigation of innovation systems must be conducted at the regional level as well as the comparatively well-documented national level. They go on to say that research should move beyond the study of successful regions, which have been the major focus of studies in industrial geography.

The chapter proceeds by analysing the factors and processes which operate within and beyond the Møre and Romsdal administrative region in Norway. They map the innovation infrastructure that supports the region's innovative capacity and look closely at what kinds of networks of local institutions and local firms form the regional innovation system.

In Part III, the strategies adopted by governments for pheripheral regions are examined. In Chapter 6, Robert Huggins examines the impact of increased networking awareness upon new models of regional development and competitiveness. This is done with particular regard to technology policy. He investigates the role of networks, both of a human and electronic nature, with respect to SMEs who are often dependent on information provision through external sources. In this context, he examines questions about the effects of networking on regional innovation strategies and the promotion of inter-firm collaboration.

The chapter focuses on the role that institutions such as 'technopoles' can play in the industrial regional economy and what their most appropriate characteristics might be. This is accomplished by a detailed study of a model that is undergoing a feasibility study in Wales by designing a 'network of networks'. This offers opportunities for information exchange at both the regional and global levels. The 'South Wales Technopole' project is a European Union (EU) SPRINT-funded study being undertaken by the Centre for Advanced Studies in the Social Sciences at the University of Wales in collaboration with the Welsh Development Agency.

The Welsh feasibility study is compared with the models of regional development that exist in Denmark and North-Rhine Westphalia. These have based competitiveness and modernisation strategies on innovation support networks and the increased participation of their respective SMEs. These comparisons indicate pointers to successful networking and increased competitiveness in peripheral regions.

In Chapter 7, Andy Pratt offers an account of the changing shape and form of innovation networks and institutions in developed industrial economies. He argues that, to date, this debate has been dominated by discussions of grand transitions' from Fordist mass-production to post-Fordist batch production. Associated with this has been a concern with the social and economic contexts that are perceived to be necessary, or sufficient, to support, or promote, economic development. Researchers have highlighted the role of the institutional and network structures within which firms operate. A further dimension of the debate has had an epistemological character implicating either macro-structures (regulation theory) or micro-actors (flexible specialisation) in explanations of the transition process.

This chapter attempts to cut across these debates in a fashion that is sceptical of the explanatory power, and extent of the applicability, of grand transition theories. It accepts that networks and institutions have always been important in industrial development; it is their exact nature, form and effect that are in question.

The chapter argues that the contemporary discussion about networks and institutions ignores the question of power. In order to understand the consequences and effects of different forms of economic arrangements for various actors and collectives, analyses of the relations between power and institutions are clearly required.

Chapters 8 and 9 look at the very significant attempts, inspired by the Ministry of International Trade and Industry (MITI) in Japan, to develop their peripheral regions by means of their 'technopolis' strategy. In Chapter 8, Sang-Chul Park describes the perceived problems confronting the Japanese at the beginning of the 1980s. These included: trade conflicts with their major trading partners, the opening up of their domestic markets and the need for basic research. These challenges were met by MITI with the launch of its 'New Industry Plan' at the beginning of the decade.

The New Industry Plan is made up of six linked strategies. These are: joint R&D projects, strategic alliances with foreign countries, the technopolis plan, telecommunication networks, venture capital and business, and the promotion of selective imports. Among these strategies, the Technopolis Plan plays a significant role in carrying out the decentralisation of industry and co-ordination between local governments, industries and academics. One of its main aims is to contribute to the future development of peripheral regions. This is to be achieved by dispersing the locations of new applications and combinations of existing technologies, assisting the emergence of new applications and technologies, and preparing for major technological innovations after the 1990s.

Sang-Chul Park focuses on how the Technopolis Plan will affect regional development and how this may be achieved. He also explores how technology transfer and the creation of new employment will take place in the technopolis regions.

In Chapter 9, Rolf Sternberg presents some original information on the effects of the technopolis policy so far. He analyses the economic and technological development of the so-called 'Silicon Island' of Kyushu in southern Japan. He investigates the reasons for the growth of high-technology industries in this part of the country, which used to be dominated by primary sector activities until the 1980s. He also analyses the impact of the technopolis programme on the regional development of Kysushu. He uses new data on all 26 technopolis sites to examine whether the technological disparities between the metropolitan core of Tokyo-Osaka-Nagoya and peripheral areas like Kyushu have been diminished as a result of the technopolis programme. The results of this analysis have important implications for European policy makers – such as those involved in French technopole development.

One of the most important requirements for state-of-the-art regional development is the transfer of the relevant and most up-to-date technology from its original sources to firms and institutions located elsewhere. There has been much discussion about how this may be achieved. Often, this debate fails to make the important conceptual distinction between information and knowledge. Among other differences, information may be transmitted electronically while knowledge and active understanding reside in the heads of individuals. The differences and alternative policy strategies that this important distinction lead to are discussed in Part IV.

In Chapter 10, Nic Komninos presents some recent developments in technology transfer theory and policy. He argues in favour of diffused and network strategies of technology transfer.

In some European countries, as in Japan, technopoles have been an important strategy for innovation and technology transfer. They have provided public R&D, start-up finance, consulting, marketing and other services to firms. Nevertheless, after two major waves of technopolitan development in Europe, between 1969 and 1973 and from 1983 to 1993, some major disadvantages in their technology transfer have become apparent. These include their degrees of localisation and the lack of external economic linkages.

After developing these criticisms, Nic Komninos goes on to discuss some recent developments in technology transfer theory and policy. These include the EU SPRINT programme, the new programme for telematic services of the fourth R&D programme, and the Bangemann report on information services and infrastructures.

These proposals develop new concepts for innovation and technology. They are characterised by a quasi-activist approach, with emphasis on networking and institutional external economies. Finally, Nic Komninos argues that there is a need for diffused strategies of technology transfer. These would involve institutions and infrastructures for multi- and non-centrally organised networks which create technology transfer links directly among firms.

This discussion leads on to Chapter 11, in which Keith Tanner and David Gibbs examine local authority strategies for providing and using information and communication technologies (ICTs) in local economic development. They outline the nature, diversity and uneven spatial distribution of current strategies that have been adopted by local authorities in Britain. Their analysis is based on the results of a questionnaire survey of some 200 metropolitan, county and district authorities. These represent all regions and both urban and rural areas.

The local initiatives are classified in terms of their overall aims, which provide details of the spatial variation of ICT policies and initiatives in Britain. A number of problem areas are also identified, which hinder the ability of local authorities to develop strategies. These include: lack of finance, unqualified personnel and the absence of national co-ordination.

Further research reveals several policy issues which give cause for concern. In particular, one major criticism is that both prior and *post hoc* evaluations may be inadequate or misdirected. The chapter concludes with an analysis of the ICT policies and initiatives that have developed in the Manchester area, to highlight the policy issues involved.

In Chapter 12, Helen Lawton Smith examines the movement of people from national laboratories in the UK, France and Belgium as a medium for knowledge and technology transfer. She contends that the dynamics of labour markets as the key to economic development are re-emerging as major academic and political themes and quotes the 1993 White Paper *Realising Our Potential* in support of this notion.

Her argument is that it is becoming increasingly apparent that mobility of personnel is a critical issue in the management of research. This is because technology transfer and the movement of scientists and engineers are part of the same process. People transfer involves the relocation of technical knowledge which, while benefiting recipient institutions in particular places directly, also feeds into existing networks of contacts. This is increasingly important because of changing technological imperatives due to first, convergences in technologies such as telecoms, television and computers; and second, the need to combine scientific and engineering knowledge located in other institutions inside and outside national boundaries. In this scenario, efficient networking becomes essential.

The chapter focuses on scientific labour markets in Europe and, in particular, on the contribution made by case study national laboratories in the UK, France and Belgium. National laboratories are defined as those which operate under the aegis of public authorities – even though they may not be directly funded from the public purse. The important feature is that their function is determined by a central or regional government department.

Helen Lawton-Smith argues that the move to commercialisation of public sector science in the UK has two obvious effects: the first is that it is leading to a declining resource base in science and engineering skills in the UK, and the second is that national laboratories in France and Belgium have been much more active in promoting mobility, and in the creation of new networks in which knowledge is transferred, than those in the UK. The consequence of the latter is that this achieves a more flexible, and potentially more successful, approach to innovation which integrates skills from outside and inside domestic economies.

The contributions to this volume identify a number of key issues with respect to innovation and the regional question. The first of these is the need to understand the special roles of core metropolitan regions with respect to national innovation. The evidence shows that on the one hand they tend to be the most innovative geographic concentrations and on the other hand they show few, if any, of the characteristics hypothesised in network and new industrial district theories. They are not the locations of collaborative supply-side networks. More often they contain secretive and competitive firms whose networks are with demand-side export customers. The ability to produce local actions which make them competitive in global markets is an important indication of their innovative success.

The second key issue is that the more peripheral western regions are all interested in overcoming their peripherality by developing local supply-side networks and information technologies. In doing so they may be misunderstanding what makes more central regions successful. One of the key features of the latter is the concentration of highly qualified professional workers. It is mostly such people who learn and innovate. Regions, networks, information technologies and new industrial districts are inanimate objects which, by themselves, can do neither of these two things. Even highly qualified professional workers are limited in their abilities to innovate, without adequate public and private funding. The Japanese technopolis programme aims to bring all these ingredients together in many of the country's more peripheral regions. It is shown that massive, and very long-term, efforts are required to overcome the innovative advantages enjoyed by the central core megalopolis.

The third and final major issue concerns the importance of technology transfer and diffusion and how best to achieve it. An important distinction must be made here between information and knowledge. Much of the evidence shows that policy makers often proceed on the basis that information transfer by technological means is sufficient to bring about significant technology transfer. This is seldom the case. Knowledge contained in highly qualified professional workers' heads is an essential ingredient of real technology transfer. Thus attempts to generate spin-offs by encouraging staff to move out of government funded research establishments may be a better example of real technology transfer.

PART II

Core Metropolitan Regions

CHAPTER 2

The Origins and Characteristics of Innovation in Highly Innovative Areas
The Case of Hertfordshire

James Simmie

Introduction

This chapter rehearses some of the basic theoretical positions which inform the analyses to come. These start with economic arguments based on the insights provided by Joseph Schumpeter. This is followed by a discussion of the critical issue of globalisation. This is argued to be both a major post-war development and a significant issue for local economies.

The counter arguments concerning the importance of localities for innovation are also reviewed. These include both the ideas of flexible specialisation and new Marshallian industrial districts. Both these ideas are argued to be place-specific and limited to industrial sectors which are seeking to extend old product life-cycles.

The role of localities with respect to innovation is argued to be more akin to the key conditions identified by Porter (1990) in connection with the competitiveness of nation states. Thus, local economies are argued to be facing both inwards, in developing their own innovative capacities, and outwards, to compete in global markets. Innovative products are said to be an essential feature of international competitiveness. In these circumstances, local economies and their firms are said to operate at the global/local interface.

Award-winning innovations in Hertfordshire, one of the UK's most high-tech counties, are used to explore these hypotheses in more detail. Some preliminary findings of an empirical survey are presented. These suggest that innovation in core metropolitan areas is a more chaotic activity than some of the available theories suggest. There is not much evidence of either systematic networking or

---

The author would like to acknowledge the support of the ESRC grant number R000221536 for this study.

high-level linkages contributing to the award winning innovations. Government regulations do not appear to contribute to innovation. Thus, theories that include networking, local industrial organisation or regulatory regimes do not appear to fit the Hertfordshire case.

The New Space Economy

Schumpeterian Economics and the Role of Space

The role of space in most Schumpeterian economics is incidental to the major concern with innovation and technological development. Nevertheless, some Schumpeterian analyses do have significant spatial implications. Practical innovations and technological developments take place somewhere and not in a placeless vacuum. The alternative considerations raised by the Schumpeter 1 and 2 models, with respect to the roles of small and large firms in innovation, have given rise to a series of simplified dualisms in subsequent analyses, which are often tied to alternative spatial scenarios. These have been summarised by Gordon (1991).

Many of these contrasting dualisms reflect the Schumpeterian belief that we are living in times of change. The 1980s and 1990s could be the the depression phase of the fourth Kondratieff long wave and the 2000s the beginning of the recovery of the next wave. As such, history may eventually show our experiences to have been those of moving from one epoch to another.

Although it is difficult to perceive exactly what will emerge as the major characteristics of the next epoch, many of the posited alternatives are concerned both with the organisational forms and the spatial arrangements of innovations and high-technology as they drive contemporary change. They include: pervasive concerns with the relative roles of large multinational corporations as compared with smaller firms and the causes of spatial agglomeration of innovative economic activities.

One side of the argument represented, for example, by Froebel, Heinrichs and Kreye (1980), Henderson and Castells (1987), Amin and Robins (1991) and Thrift (1989) is that a global economy has developed which is dominated by large multinational corporations (MNCs). The decisions of these MNCs, on where they conduct such activities as R&D and production, determine to a large extent what economic activities agglomerate in particular places.

The other side of the argument represented by Piore and Sabel (1984), Porter (1990), Lundvall (1992) and the new Marshallians – such as Scott and Storper (1987), Storper and Christopherson (1987) – is that local places are becoming more, not less, important in their contribution to innovation and high-technology. The focus of consequent research is quite different, according to which side of this global/local divide is taken. The main propositions of the two views are outlined briefly below.

Globalisation

There can be little doubt that capital is concentrating and centralising at the level of the international economy. The corporate vehicles for this concentration are the multinational corporations (MNCs) with control centralised in their respective headquarters, which are often located in and around world cities – such as London, Tokyo and New York.

There is plenty of data which confirms the significance of MNCs as major shapers of the world economy. Even by 1980, for example, only 350 of the largest of them controlled economic resources which were equivalent to more than a quarter of the combined Gross Domestic Products (GDPs) of all the developed and less-developed countries put together. Somewhere between 25% and 40% of all world trade consists of purely internal transfers between the subsidiaries of MNCs (Sutcliffe 1984). High-technology firms are also some of the major players in the global economy. Characteristically they have low levels of forward linkages. This tends to confirm the findings of several researchers: that high technology firms operate in global markets.

The argument developed following these kinds of data is that localities, regions and even national territories are being re-shaped according to the global economy and its main players the MNCs. Following this formulation, there is a spatial division of labour and a spatial division of innovation. The large enterprise is able to split its activities into units and to localise and disperse these units in the most favourable places in terms of work and industrial culture (Massey 1984; Aydalot 1979). Regions at the nodes of the global network have a large autonomy. The further they lie from this central node, the more regions are locked into the international division of labour and resemble the old Fordist branch centres (Amin and Robins 1990a).

These views lead to trenchant criticism of the argument that regions and localities are growing, rather than declining, in independence and importance in the contemporary innovative and high-technology era. It is argued, for example, that MNCs, with their global networks, have far more impact on the world economy than locally-embedded firms. Therefore, to an MNC, flexibility is a matter of industrial organisation on a global rather than a local scale. As far as they are concerned, the issue is not how to increase local area autonomy but how to create more efficient corporate integration. This makes industrial geography a series of maps of places with different roles in the international division of labour. As a result, local places experience different degrees of economic well-being and local production synergies (Amin 1991; Amin and Robins 1990b; Praat 1991).

The main problem with this argument is that places are assumed to be homogenous. Although they may be grouped into different categories such as centre and periphery, it is assumed that these are so similar as to make MNCs entirely indifferent as to which of the appropriate categories of space they allocate their appropriately divided labour. It is, therefore, assumed that a high-technology MNC like IBM is indifferent in its choice of location of R&D as between places like New Jersey, Dortmund or Sidney; or of production of personal computers as between Sonoma, Taiwan or even Patagonia. In practice this is clearly not the case. Differences between localities within and between different categories of place are significant and do matter. This appears to be true, even within the acknowledged context of the globalisation of the world economy.

The questions that still need to be answered are, therefore, given the globalisation of the world economy, why does location still matter? And what are the characteristics that distinguish 'successful' innovative and high-technology areas from those that are not? Two of the best-known 1980s answers to these questions are examined below.

Flexible Specialisation and Marshallian Industrial Districts

One of the most influential analyses in the 1980s of why place could still matter with respect to innovative industries is the idea of flexible specialisation pioneered in the seminal work by Piore and Sabel (1984) *The Second Industrial Divide*. In it they argue that the widespread beliefs that firms are independent entities, and that small firms are linked in competitive markets whereas large firms are organised as oligopolistic hierarchies constituting entire industries, are neither an exhaustive nor accurate description of current configurations. They further argued that firms, particularly those organised in industry-embracing hierarchies, are saturating markets with traditional, standardised, mass-produced goods. As a result, consumers are demanding more specialised and differentiated goods – which mass-production systems, typically, cannot supply. The response of some firms to these changed circumstances is argued to be the development of flexible specialisation. This is a strategy of permanent innovation: firms accommodate ceaseless change, rather than try to control it. The strategy is based on flexible, multi-use equipment, skilled workers and the creation, through politics, of an industrial community that restricts the forms of competition to those favouring innovation. It is also argued that the spread of flexible specialisation amounts to a revival of craft forms of production that were marginalised during the first industrial divide, which is usually referred to as the industrial revolution.

Piore and Sabel (1984) also argue that the spread of flexible specialisation represents such a major and pervasive change that it constitutes a shift of technological paradigm. They cite examples of the re-invigoration of craft-based industries in Italy, Germany and Japan in support of the pervasiveness of what they argue to be a new paradigm. Areas based on small, craft firms in places like central and north-western Italy, Mondragon in the Basque Region of Spain (Stohr 1986) and the high fashion areas of Paris (Storper 1993) have been studied intensively to illustrate the main characteristics of flexible specialisation.

There are five major problems with the flexible specialisation thesis in its original formulation:

1. Far from indicating a new, emerging paradigm for the industries of the fifth long wave, the examples cited usually refer to old industries surviving from previous times. Many of these industries have reached the latter stages of their product life-cycles and are mostly attempting to extend them into artificially differentiated niche markets in order to survive. Designer ceramic tiles from Italy and high-fashion from France are not the industries of a new techno-economic paradigm of the future. They also suffer from the problem that much of their niche market value is derived from such intangible assets as designer names. The products themselves can often be copied and sold for less.
2. There is not much evidence that the industries of the next innovative technological trajectory will be in any way craft-based. Dosi *et al.* (1988, p.52) suggest that they will be some combination of computers, electronic capital goods, software, telecommunications equipment, optical fibres, robotics, flexible manufacturing systems (FMS), ceramics (not tiles), data banks, information services, digital communications networks and satellites. It is hard to see how any of these industries could be organised on a craft basis and restricted to particular geographic areas.
3. The flexible specialisation theory ignores the growing globalisation of the world economy. Far from a general increase in the importance of local groups of small firms, many writers have argued that what we are witnessing is a deepening of the historical trends towards global integration of local and national economies and the international centralisation of command and control. The dominant, though not uncontested, tendency is towards market homgenisation, industry globalisation and firm integration (e.g. Doz 1987).
4. Where they actually exist, the networks that are so important to the flexible specialisation thesis really indicate a deepening and extension of the structure of oligopolistic behaviour and control. What is at work is not fragmentation, decentralisation and increasing organisational autonomy, but more effective corporate integration across vertical, horizontal and territorial boundaries (Amin and Robins 1991).
5. There do not appear to be many true examples of flexible specialisation in practice. While craft communities are to be found in the Third Italy, some of the other frequently cited examples of the genre, on closer examination, do not seem to support the thesis. Silicon Valley in California, Route 128 around Boston and Silicon Glen in Scotland, for example, are not only all significantly different from one another, but also display very few of the characteristics of flexible specialisation.

Despite these criticisms the idea of flexible specialisation has been combined with that of Marshallian industrial districts in order to seek to 'explain' the undoubted emergence of innovative, high-technology agglomerations in certain places. This spatially-focused analysis is examined next.

Marshallian Districts

The idea and characteristics of spatially concentrated industries is associated with the economist Alfred Marshall. He coined the phrase 'industrial district' in 1890 to describe such areas. He subsequently developed the idea that 'The leadership in a special industry, which a district derives from an industrial *atmosphere* (emphasis added), such as that of Sheffield or Solingen, has shown more vitality than might have seemed probable in view of the incessant changes of technique. It is to be remembered that a man can generally pass easily from one machine to another, but that the manual handling of a material often requires a fine skill that is not easily acquired in the middle age: for that is characteristic of a special industrial atmosphere. Yet history shows that a strong centre of specialised industry often attracts much new shrewd energy to supplement that of native origin, and is thus able to expand and maintain its lead' (Marshall 1919, p.287).

The idea was taken up and re-invigorated by Becattini (1990). Studies, originally inspired by the idea of flexible specialisation in fast-growing industries such as textiles, footwear and ceramic tiles in the Third Italy, claimed to have rediscovered industrial districts in the areas specialising in these industries. It has also been argued that some high-tech industrial complexes in California operate as industrial districts (Saxenian 1991; Scott 1993).

Critics of the concept of Marshallian industrial districts point to the rarity of some of their claimed characteristics in modern economies. Thus, local allegiance, co-operation, trust relations and social and institutional solidarity are hard to find. At any rate, they do not appear to be common in high-technology clusters in the US. If anything, they have become increasingly rare in the developed economies and are difficult to nurture in places where they do not already exist.

Analyses of industrial districts tend to ignore the significance and effects of the global economy as far as different areas are concerned. Attention is focused on the internal social, political and institutional characteristics of areas identified as industrial districts. As a result, there is a tendency to treat them as industrial islands as much as districts. But, as Storper (1993) remarks, the main characteristics of technology districts should be seen in the context of the principal trends in the international economy. They are national industrial areas of specialisation based on often unstable technologies which are being urged to combine and become even more unstable by the constant development of scientific research and markets.

Although some areas can be readily distinguished as potential neo-Marshallian industrial districts, they can also be seen to embrace a wide variety of forms and characteristics. There is little in the way of explanation in the industrial district concept of why such areas arise in the first place, the variety of types that emerge, and what are the functional relationships between industrial imperatives and spatial form.

While on the one hand it is descriptively and empirically the case that something like industrial districts can be identified on the ground, on the other hand the recent redevelopment of neo-Marshallian thought on the subject has not offered much in the way of satisfactory explanation of their existence, variety, characteristics, and potential links between industrial decisions and spatial results. So, while it is easy to agree that innovative, high-technology agglomerations can be found in many countries, it is not so easy to explain why.

In order to advance such an explanation, it is argued here that it is productive to start with the insights provided by Porter (1990) in *The Competitive Advantage of Nations.*

Nation States and Local States

In his influential book, Porter argues that

> 'Competitive advantage is created and sustained through a highly localized process. Differences in national economic structures, values, cultures, institutions, and histories contribute profoundly to competitive success. The role of the home nation seems to be as strong as ever. While globalization of competition might appear to make the nation less important, instead it seems to make it more so. With fewer impediments to trade to shelter uncompetitive firms and industries, the home nation takes on growing significance because it is the source of the skills and technology that underpin competitive advantage'.(p.19)

While it is clearly true that the economic autonomy of nation states is constrained by the actions of MNCs, nations continue to play significant roles in the conditions under which MNCs operate. Political boundaries create one of the most important ways in which location specific factors are packaged (Thrift 1989, p. 149). They create discontinuities in conditions of supply and demand. Governments can help both to create and destroy the competitive advantages of the firms or elements of MNCs which operate within their boundaries. According to Thrift, such national differences could be the single most important factor in creating global shifts in economic activity.

Porter (1990) identifies four major characteristics which differentiate between national and regional politico-administrative places:

1. Factor conditions.
2. Demand conditions.
3. Related and supporting industries.
4. Firm strategy, structure and rivalry.

Each of these can be influenced in various ways by the political units in whose particular territories firms, or parts of firms, seek to operate. While most of them cannot escape international competition in the global economy, the politically created conditions under which they operate can make a significant difference to their success or failure.

Porter also argues that there is an association between vigorous domestic or regional rivalry, in technological terms, and the creation and persistence of competitive advantage in an industry. The local operating environment of firms can play an important role in, for example, the diffusion of new product and process technologies. Geographic concentration, even of rival firms, enhances the benefits of strong competition. This is because it:

- stimulates a fast diffusion of new technologies
- helps upgrading suppliers through competition and intensive co-operation with customers on R&D
- puts pressure on political support in creating specialised factors such as specific training and research centres
- stimulates firms to fund local training and research centres themselves.

The main thrust of these arguments is that, even in the context of the global economy, space does matter because it is divided up into political trading blocks and nation states. These political entities can, and do, make significant differences to the local economic environment in which firms have to operate. These differences are so significant as to have major impacts on the processes of economic globalisation themselves, the competitive success or failure of firms and of entire nation states.

These arguments have been taken further to apply not just to whole nation states but also to smaller political entities within them. Politically distinct locations, such as regions or even sub-regions, can influence the competitive conditions under which firms operate. In so doing, they also effect the competitive success or failure of the local state as a whole.

Porter's work, therefore, leads to interesting research questions about what local, politically-created conditions contribute to the international competitive success of firms in the area. It links the globalisation of the economy with the local conditions which make firms competitive in the international arena. In this analysis it focuses attention on what national and local states can do with respect to factor conditions, demand conditions, related and supporting industries, and firm strategy, structure and rivalry in order to contribute to the international competitiveness of locally operating firms. These relationships form, what is called here, the global/local interface.

The Global/Local Interface: An Explanatory Framework

So far it has been argued that innovation is an important element determining the national and international competitive performance of firms. It has also been argued that the external relationships of firms both with their local production spaces and their global market places, termed the global/local interface (GLI), are highly significant in their generation of different types of innovation and their competitive performance. In principle, the multiple considerations and decisions that different firms take, to position themselves as advantageously as possible with respect to their individual GLIs, could lead to a wide variety of locational choices. In practice this is not the case. There are a limited number of well documented types of spatially clustered innovative high-technology agglomerations.

The starting point here is a desire to understand and explain the underlying forces shaping the characteristics of particular, innovative, high-technology local space economies. The reasons for focusing on innovation and high-technology have been outlined above. They relate primarily to the neo-Schumpeterian view that innovations play a major role in driving economic growth; and that high-technology innovations are key phenomena in each successive techno-economic paradigm.

It is argued that in order to achieve such an understanding, it is necessary to adopt a theoretical approach which may combine, in a consistent way, the seemingly irreconcilable conclusions that, on the one hand, local space economies are dominated by global forces, while on the other hand, some of them are becoming more important in terms of their contributions to competitive economic growth based on innovation and high-technology. The theoretical proposition which seeks to make this reconciliation here is the global/local interface.

The global/local interface can be described as the nexus between the forces which increasingly expose more firms to international, competitive markets and the structures and strategies which they adopt to compete successfully in those markets; and the local conditions, regulations and regimes which enable parts of, or whole, firms to export competitively into those international markets.

It is argued that the globalisation of the world economy is a growing fact of economic life. This is both partly a cause and partly an effect of the growth of MNCs. It is also greatly enabled by the development of modern forms of transportation, communications and information technology. It is not only the result of the actions of MNCs. Many modern, innovative high-technology products such as whole-body scanners, and services such as software design, are provided by relatively small companies that can only recoup their original development costs if they sell in large enough numbers on large world rather than small national markets.

It is also argued that the activities of all innovating high-technology firms take place somewhere in particular places. This raises the questions of why firms select some places rather than others, and what conditions in the selected places contribute to innovation which is internationally competitive. It is descriptively clear that some forms of innovative industrial districts exist. Descriptions of them are also notable for the variety of different characteristics that they describe. Only a few of them are especially characterised by flexible specialisation. An analysis is therefore required which can embrace a variety of different types of innovative areas and begin to explain the different reasons for their relative success or failure in the global economy.

In so far as neo-Schumpeterians have been concerned with the roles of local conditions in promoting internationally competitive innovations, they have argued, as mentioned above, that there is a tendency for the more radical innovations to occur in metropolitan areas. Some have then gone on to use an 'epidemic-hierarchic' model to describe the subsequent diffusion of innovations down through more minor nodes in international and national urban hierarchies. This does not explain what it is about metropolitan areas that is particularly conducive to innovation or what limitations are present in other areas.

One general proposition, which begins to address the Schumpeterian problem of uncertainty, is that knowledge and information are key factors in modern high-technology innovation and that some locations, such as metropolitan areas, provide greater and quicker access to a wider range of knowledge and information. This tends to alleviate the uncertainty problems of firms operating in such places. Rapid generation, access and understanding of relevant knowledge and information, as a result of differentiated localised processes, could be a major factor in sustaining international competitive advantages.

This raises the question of what might constitute relevant knowledge and information in the context of internationally competitive high-technology innovation. In addition to the more obvious requirement for high-level scientific and technical knowledge, it is instructive to consider the four major characteristics which Porter (1990) argues provide some nations or places with international competitive advantages. These have been mentioned above and are:

1. Factor conditions.
2. Demand conditions.
3. Related and supporting industries.
4. Firm strategy, structure and rivalry.

In the context of seeking to understand the local spatial GLIs which promote and enable competitive high-technology innovations, it is argued that these four elements are highly significant, have different balances of global and local relevance, and make different contributions to the export potential and capabilities of local firms. Furthermore, the differences between them in different areas may explain the variety and differences which have been described in different high-technology areas.

It is argued here that innovative and internationally competitive high-technology firms require knowledge and information about the four elements listed above in order to reduce their present and future degrees of uncertainty in becoming, and remaining, competitive in national and international export markets. In terms of their respective GLIs, they require knowledge and information in order to act successfully.

The Hertfordshire Case Study

The selection of Hertfordshire as an illustrative case study of these theoretical arguments is based on two major spatial considerations. First, it forms part of an arc of counties located around the west of London which, together with London itself, contains higher absolute and relative concentrations of high-technology and innovative industries than anywhere else in the UK (see, e.g. Hall, Breheny, McQuaid and Hart 1987; Castells and Hall 1994).

Table 2.1 shows that Hertfordshire had the highest relative concentration (Location quotient 3.60) of high-technology employment of any county in Great Britain (GB). It also had the second highest absolute numbers of such workers after Greater London in 1981. Table 2.1 also shows that Greater London, together with its western arc also contained around one-third of all high-technology employment in GB in 1981.

Table 2.1 Concentrations of high-technology industry, by county groupings, Great Britain 1981

	Employment	*Location quotient*
London western arc		
Greater London	91,400	0.85
Hertfordshire	45,100	3.60
Hampshire	33,800	2.00
Berkshire	19,700	2.04
Buckinghamshire	6,900	1.15
Surrey	18,200	1.79
Total	215,100	
North-west England		
Greater Manchester	27,300	0.86
Lancashire	30,000	2.01
Total	57,300	
Central Scotland (Silicon Glen)		
Strathclyde	23,600	0.88
Lothians	9,000	0.93
Fife	8,100	2.24
Total	40,700	
Total three areas	**313,100**	
Great Britain	**640,900**	

Source: Castells and Hall 1994, p.147

Second, towns in Hertfordshire are also the locations for some of the highest absolute and relative concentrations of research and development employment. In so far as the presence of such employment is an accurate indicator of local innovative capacity, Hertfordshire stands out among British counties. In 1991, for example, two Hertfordshire towns, Welwyn (2,642) and Hertford and Ware (1,794), were fifth and eighth respectively in Britain's top ten areas with the highest absolute concentrations of R&D employment (Howells 1995).

Table 2.2 shows that the towns of Hertford and Ware (LQ 17.62) and Welwyn Garden City (LQ 12.35) ranked third and fourth respectively in terms of the relative concentration of R&D employment in British cities in 1991. This ranks them above the much vaunted towns of Bracknell in the M4 corridor, and Cambridge with its university and science park.

Table 2.2 Top ten areas with highest relative concentrations of R&D employment in Britain 1991

Local labour market area		*Location quotient*
1.	Didcot	46.02
2.	Thurso	20.54
3.	Hertford and Ware	17.62
4.	Welwyn	12.35
5.	Huntingdon	9.95
6.	Bracknell	9.59
7.	Salisbury	9.29
8.	Weymouth	7.73
9.	Malvern	7.55
10.	Cambridge	7.55

Source: Howells, J. (1995)

According to Porter (1990), it is an area's industries that make it internationally competitive. The argument advanced here is that, in turn, it is innovation that makes those industries themselves competitive. It is further argued that using an iterative, as opposed to a linear, model of innovation means that consideration should be given to both technology push and demand pull as drivers of innovation.

Even in Hertfordshire, not all firms and organisations are innovative or successful at claiming significant shares of global innovative markets. In order to have a reasonable chance of identifying which firms have been both innovative and commercially successful, it was decided to investigate only those firms and organisations which had recently won awards for technological advances, export success and innovation.

There are some eight UK and a further six European awards for innovation. These support a range of activities from pre-competitive research, through innovative collaborations, to information provision and strategic programmes in key activities. Two UK programmes were selected because of their combined focus on export market success, technological advance and innovation. They also covered a range of firm sizes from very small to very large. The programmes selected were: the Queen's awards to industry for export and technological success and the Small Firm Merit Awards for Research and Technology (SMART).

The Queen's award for industry was started in 1966. It was divided into two separate awards for export and technological achievement in 1975. A further category for environmental achievement was introduced in 1992.

The Small Firm Merit Awards for Research and Technology (SMART) is a UK nation-wide programme to support single company innovation projects. The scheme was started in 1988. It involves a two-stage annual competition for small companies, defined as those employing up to 50 people, who want to start an innovative project but have been unable to raise initial funding.

The total award-winning sample frame used in this study is shown in Table 2.3. Taken together, it covers innovation in a wide range of different sized firms. It also includes innovations that are the result of both technology push and demand pull.

Table 2.3 Innovation award winners in Hertfordshire 1985–1995

Awards	*Numbers*	*Percentage of sample*
1985–1995		
Queen's award for exports	23	40
Queen's award for technology	9	16
Queen's award for both	2	3
1988–1995		
SMART award	24	41
Total	**58**	**100**

The detailed analysis in Table 2.4 shows that the towns where the most awards have been won are Hemel Hemstead (8), Royston (6), St. Albans (5), Stevenage (5), Watford (4), Baldock (3), Harpenden (3), Hatfield (3), and Hertford (3). These are a mixture of old, new and expanded towns. They are medium sized in terms of population and are located throughout Hertfordshire.

Early Results

At the time of writing, about half of the total respondents identified as winning awards have been interviewed. The results discussed here are, therefore, early and preliminary. Nevertheless, they provide some useful indicators of potential research findings.

Table 2.4 Innovation award-winning towns in Hertfordshire

Town	*Award*				
	Queen's Export	*Queen's Tech.*	*Queen's both*	*SMART*	*Total*
Hemel Hemstead	7	1	–	–	8
Royston	1	1	–	4	6
St. Albans	1	1	–	3	5
Stevenage	2	1	1	1	5
Watford	2	1	–	1	4
Baldock	–	–	–	3	3
Harpenden	2	1	–	–	3
Hatfield	–	–	–	3	3
Hertford	1	–	1	1	3
Borehamwood	–	1	–	1	2
Hitchin	1	–	–	1	2
Hoddesdon	2	–	–	–	2
Kings Langley	–	1	–	1	2
Welwyn Garden City	–	–	–	2	2
Berkhamsted	–	1	–	–	1
Bishops Stortford	1	–	–	–	1
Chorleywood	–	–	–	1	1
Letchworth	1	–	–	–	1
London Colney	1	–	–	–	1
Potters Bar	1	–	–	–	1
Rickmansworth	–	–	–	1	1
Ware	–	–	–	1	1
Totals	**23**	**9**	**2**	**24**	**58**

First, a list of significant innovations discovered so far indicates a fairly chaotic bubbling-up of widely differing innovations in different activity headings. Table 2.5 shows that these vary from military gas detectors, through medical diagnostics for glaucoma, to a new catalytic converter coating. These findings indicate that innovation may be a more chaotic and complex phenomenon than some previous theories suggest. Propositions that suggest systematic regularities may be losing something of the reality of innovation in practice.

Second, in so far as some general properties do emerge in the analysis contained in Table 2.5, they indicate the fairly consistent importance of high-level knowledge as a basis for innovation. In the examples cited, this knowledge includes physics, chemistry, engineering and computing. Many of the innovations are specific, often niche or customised, applications of such knowledge.

Third, there are indications of both technology push and demand pull in the innovations discovered so far. Technology push includes state of the art product technology, miniaturisation, leading-edge software and new measurement techniques. Demand pulls include the opening up of new markets and increased international competition.

Table 2.5 Main winning innovations

Innovations	*Activity headings*
Design and supply of insulation materials and services within oil, gas, power and petrochemical industries	2437
Electronic page composition system	3276
New technology for hand-held military gas detectors	3290
New production method for credit card manufacturing equipment	3302
Multimedia video and graphics boards for personal computers	3444
Fourth generation software for commercial applications, fast hardware system suitable for communications and military applications	3454
New catalytic converter coating	3530
Software for increasing speed of electrocardiogram analysis	3720
Confocal microscopes	3732
Exports of magazines on computing, cartoons, comics to the Middle East	4752
Export of computer system of hospital design called 'Nucleus'	5010
Diagnostics for glaucoma	8370
Photostable synthetic pyrethroids (safe and active insecticides)	9400

The analysis of these innovations focuses on the contributions of the four main elements of Porter's diamond to their development and commercial success. With respect to local factor conditions, it was hypothesised that innovative actions would be facilitated in local areas by:

- reasonable-cost land and buildings
- good externality characteristics such as infrastructure, communications and environmental conditions
- local labour markets containing numbers of adequately educated and trained labour at reasonable prices

- venture and large-scale capital for initial start-up companies and to finance long-term research and development.

The interviews conducted with innovative firms so far indicate that the most important of these local factors are high-quality human resources, knowledge and information combined with the availability of capital resources. Hertfordshire provides these requirements in terms of a highly-educated workforce, which brings with it knowledge and the ability to translate information into innovative action.

Hertfordshire is also what might be described as an 'accessible rural area' with respect to London. The metropolis is the major UK centre for venture and large-scale capital markets. These can be tapped easily and on a face-to-face basis. Numerous producer services have also moved to, or started up in, Hertfordshire itself. Local factor conditions, therefore, combine knowledgeable professional labour with access to capital resources on a high trust and personal basis.

None of the other hypothesised factors have been rated as significant by the firms interviewed so far. It appears that relatively high costs for land and buildings are accepted. Externality characteristics such as infrastructure and information technology are also not thought to have contributed significantly to innovation.

As far as demand conditions are concerned, it was argued that innovative capacity would require a combination of:

- national and international information on present and potential demand conditions
- the ability to identify markets and capture adequate shares of them
- the views and positions taken by firms on national and international markets will also interface with, and have feedback influences on, what they require in terms of local factor conditions.

Among the innovations examined so far, the majority of them are exported to markets outside the UK. In this majority of cases, more than three-quarters of Hertfordshire turnover was exported. This indicates both the significance of innovation for international competitiveness and the importance of demand pull in stimulating the innovations examined.

The importance of new buyers, as far as the award-winning innovations are concerned, is high. Thus, in a majority of cases, new innovations were gaining commercial success as a result of demand by new foreign buyers.

The preliminary conclusions drawn from these findings are that demand pull from global markets is a key stimulus to innovation. New and growing markets for niche and customised products appear to be an essential ingredient of the innovations studied in Hertfordshire.

Turning to the contributions of related and supporting industries, it was initially hypothesised that:

- backward linkages with suppliers can only aid international competitiveness if those suppliers are themselves technically competitive
- forward linkages with national customers only aid international competitiveness if those customers demand goods and services which are up to international standards
- producer services such as finance, marketing, advertising, transport, communications and information technology also need to be internationally competitive in order to enable high-technology firms to compete with their international rivals.

Most of these hypotheses have turned out not to be the case so far. In particular, the importance of any local networks in contributing to innovation are rated as 'low' or 'none at all' by local firms. There appears to be a general lack of interconnectedness with respect to high-level activities in Hertfordshire.

Among the few local linkages discovered so far, most appear to involve low-level support services such as office suppliers and cleaners. There are few, if any, examples of high-tech collaborations on innovative projects. These findings echo those of Decoster and Tabariés (1986) with respect to *La Cité Scientifique*, south of Paris.

Finally, with respect to firm strategy, structure and rivalry, it was hypothesised that:

- In limited circumstances, congeries of small local firms may adopt a strategy of flexible specialisation.
- In others, vertically integrated structures focusing narrowly on core business are adopted. A wide variety of strategies and structures exist between these two extremes.
- Local firm rivalry can assist competitive innovation by accelerating the circulation and diffusion of new ideas and innovations.

Initial evidence suggests that the UK economy is too small on its own to have sufficient competing companies in any particular industrial sector. Most of the innovative companies described their main competition as being located either in North America or world-wide. A small minority regarded the main competition as being located in the UK or western Europe. This tends to confirm the initial impression of the apparent chaotic nature of innovation in Hertfordshire. Thus, although the area is being used as a home base by many innovation award-winning companies, most of the other companies with whom they are in competition are located abroad. The result is that local firm rivalry does not appear to contribute much to knowledge and innovation diffusion, at least in Hertfordshire.

Conclusions

Preliminary conclusions from the study so far suggest that innovation in core metropolitan areas is a more chaotic activity than some of the available theories suggest. There is not much consistency in the the types of innovation discovered; nor is there much evidence of either systematic networking or high-level linkages contributing to the award-winning innovations. The firms interviewed so far do not regard government regulations as contributing to innovation. Thus, theories that include networking, local industrial organisation or regulatory regimes do not appear to fit the Hertfordshire case. Instead, a more fluid system is emerging from the research results.

In this context, numerous, seemingly unrelated, innovations have bubbled-up from a complex knowledge base which is embodied in the highly-educated professional workforce that has chosen to live in and around Hertfordshire. The nearby availability of venture and long-term capital emanating from sources based in London is also a major contributory factor to innovations in the area. The successful innovating firms can, and do, compete in global markets from their base in Hertfordshire. In doing so, they are in competition with other companies – mainly in North America and other world-wide locations.

The UK economy on its own does not appear to be large enough to support many competing firms in the same industrial sector. Thus, innovation diffusion is limited by the lack of other organisations working at high levels in the same industrial sectors. This also indicates that the demise of a very few firms can decimate the innovative capacity of the UK economy in a very short space of time.

Bibliography

Amin, A. (1991) 'Giant shapers and shakers of the world economy leave British hopes behind as wishful thinking.' *The Guardian*, 7 January, p.9.

Amin, A. and Robins, K. (1990a) *Industrial Districts and Local Economic Regeneration.* Paper presented at the Workshop on the Socio-Economics of Inter-Firm Cooperation, Social Science Centre, Berlin (WZB), 11–13 June 1990.

Amin, A. and Robins, K. (1990b) 'The re-emergence of regional economies? The mythical geography of flexible accumulation'. *Environment and Planning D: Society and Space 8.1*, 7–34.

Amin, A. and Robins, K. (1991) 'These are not Marshallian Times.' In R. Camagni (ed) *Innovation Networks: Spatial Perspectives.* London: Belhaven.

Aydalot, R. (1986) *Milieux Innovateurs en Europe.* Paris: GREMI.

Becattini, G. (1990) 'The Marshallian industrial district as a socio-economic notion.' In F. Pyke, G. Becattini and W. Sengenberger (eds) *Industrial Districts and Inter-firm Cooperation in Italy.* Geneva: International Institute for Labour Studies.

Castells, M. and Hall, P. (1994) *Technopoles of the World: The Making of 21st Century Industrial Complexes.* London: Routledge.

Decoster, E. and Tabariés, M. (1986) 'L'innovation dans un pôle scientifique: le cas de la Cité Scientifique Ile de France Sud.' In P. Aydalot (ed) *Milieux Innovateurs en Europe.* Paris: GREMI.

Dosi, G., Freeman, C., Nelson, R., Silverberg, G. and Soete, L. (1988) *Technical Change and Economic Theory.* London: Pinter.

Doz, Y. (1987) 'International industries: fragmentation versus globalisation.' In B. Guile and H. Brooks (eds) *Technology and Global Industry.* New York: National Academy Press.

Froebel, F., Heinrichs, J. and Kreye, O. (1980) *The New International Division of Labour.* Cambridge: Cambridge University Press.

Gordon, R. (1991) 'Innovation, industrial networks and high technology regions.' In R. Camagni (ed) *Innovation Networks: Spatial Perspectives.* London: Belhoren.

Hall, P., Breheny, M., McQuaid, R. and Hart, D. (1987) *Western Sunrise: The Genesis and Growth of Britain's Major High Technology Corridor.* London: Allen and Unwin.

Henderson, J. and Castells, M. (eds) (1987) *Global Restructuring and Territorial Development.* London: Sage.

Howells, J. (1995) *The Dynamics of Research and Development: Opportunities for Growth in the South East.* Paper presented to the SEEDS Workshop on R&D in the South East, 1 Nov 1994, London Voluntary Resource Centre.

Howells, J. (1995) *Research and Development in Hertfordshire: Knowledge County and Global Research Base.* Report to Planning and Environment Department, Hertfordshire County Council. Unpublished.

Lundvall, B.A. (ed) (1992) *National Systems of Innovation: Towards a Theory of Innovation and Interactive Learning.* London: Pinter.

Marshall, A. (1919) *Industry and Trade.* London: Macmillan.

Massey, D. (1984) *Spatial Divisions of Labour.* London: Macmillan.

Piore, M.J. and Sabel, C. (1984) *The Second Industrial Divide.* New York: Basic Books.

Porter, M.E. (1990) *The Competitive Advantage of Nations.* New York: Free Press.

Praat, H. (1991) 'Principles of networking.' In M. Schmidt and E. Wever (eds) *Complexes, Formations and Networks.* Utrecht: Nijmegen.

Saxenian, A.L. (1991) 'The origins and dynamics of production networks in Silicon Valley.' *Research Policy 20,* 423–437.

Scott, A. (1993) *Technopolis: High-Technology Industry and Regional Development in Southern California.* Berkeley: University of California Press.

Scott, A.J. and Storper, M. (1987) 'High technology industry and regional development: a theoretical critique and reconstruction.' *International Social Science Journal 112,* 215–232.

Stohr, W.B. (1986) 'Regional innovation complexes.' *Papers of the Regional Science Association 59,* 29–44.

Storper, M. (1993) 'Regional "worlds" of production: learning and innovation in the technology districts of France, Italy and the USA.' *Regional Studies 27.5,* 433–455.

Storper, M. and Christopherson, S. (1987) 'Flexible specialisation and regional industrial agglomeration: the case of the US motion picture industry.' *Annals of the Association of American Geographers 77,* 1, 104–117.

Sutcliffe, B. (1984) *Hard Times.* London: Pluto.

Thrift, N. (1989) 'The Geography of International Economic Disorder.' In R.J. Johnston and P.J. Taylor (eds) *A World in Crisis.* Oxford: Blackwell.

CHAPTER 3

The Financial and Entreprenuerial Characteristics of Innovative Small Firms in Contrasting Regional Environments of the UK

Pooran Wynarczyk, Alfred Thwaites and Peter Wynarczyk

Introduction

Uneven economic progress has characterised regional development in Britain over most of this century with some regions and sub-regions appearing to lag behind the national economy, and core regions in particular, on a more or less permanent basis. As a result, considerable academic and policy interest has been evident in this area, but the gap between rich and less favoured regions has not narrowed substantially in the UK. The search for a fuller understanding of the causes and possible solutions to these problems provides a rationale for continuing, if not increasing, research in this field. The particular focus in this chapter is upon the contribution of innovative small firms to the economic growth of regions. It has little to say about technology *per se*, but concentrates on the economic effects of innovation on the financial performance of small firms located in different regions within the UK.

Background

Why concentrate on the small and innovative firm as a subject of study? Many studies associated with local economic growth in post-war Britain concentrate on the contribution of the large and mobile firm as a source of growth. More recently, however, the role of the large firm in local development has been re-evaluated when, on average, this type of firm became associated with rationalisation and employment reduction. At the same time, the small and medium sized enterprises sector has been seen to grow quite strongly and is claimed to have re-emerged as the principal generator of new jobs and employment (Birch 1979). Although often cited as evidence of a growing enterprise culture in the UK, the majority of SMEs remain small and have a lower level of impact on employment and output than their numbers would imply. In performance terms, the SME sector is also an heterogeneous one – which is, perhaps, a result of their multifarious *raisons d'être.* For example, there are doubts surrounding the SME as a long-term solution to either national or regional economic problems. Gray (1992) points to the high rates of insolvency and business failure in SMEs as evidence of instability and the fact that the rate of VAT registrations is decreasing. From a regional development point of view, research also suggests that the birth, death, expansion and contraction of SMEs is spatially differentiated, so that some regions, particularly the core regions, seem to have benefited more from their activities than others (for a review see, for example, Elliot-White 1991).

While the messages emerging on the pros and cons of stimulating small firms in general have become somewhat confused, other researchers have devoted efforts towards identifying a subset of small firms which appear to bring the greatest advantages to a region or national economy. For example, Storey *et al.* (1989) have shown that a very small proportion of total SMEs are responsible for the majority of positive effects of the SME population on local or national development. Research in this genre has led some researchers down the routes of exploring innovative small firms as a possible set of high-growth small firms. Such firms are hypothesised to combine the two features of small-scale and technological advance, both of which are a priori expected to have a positive effect on economic advance.

Over the past twenty years, there has been a burgeoning interest in the role of technological change in economic development at a variety of spatial scales. The arguments supporting technology as an important factor in local economic development are essentially the translation down to the sub-national scale of the arguments expressed at national or global scales. The general conclusions of this debate are summarised by Malecki (1991): 'Technology is central to regional change, positive and negative, and to economic change, job-creating and job destroying' (p.7).

The importance of technological innovation in a regional development context is its ability to provide a possible foundation for new industries; for the creation, broadening and deepening of markets for regional firms by substituting new economic goods for existing and competing goods. It can also effect cost, quality and reliability. A region or country in which industrial firms achieve substantial technological progress through the generation, adaptation or adoption of new products is seen to have a competitive advantage over others making slower progress. It is perhaps not surprising, therefore, that the issue of technological progress is high on research and policy agendas and that those in lagging regions see it as a way of helping to resolve their current problems.

In support of this line of argument, Pavitt *et al.* (1987) have shown that, over time, small firms have been introducing significant new products into the UK, Rothwell (1986) has also shown that small firms are vital agents in the diffusion of technology where they take generic innovations developed elsewhere and present them in a wide variety of forms essential to meeting the expressed or latent needs of a broad range of users. SMEs are important also in introducing those incremental innovations which surround generic technologies and existing applications which broaden and deepen markets (Rosenberg 1992). In technological and market terms, the small and innovative firm could be seen as a potentially powerful force in local economic change. As most small firms, once established, continue to produce in the same locality, their technological activities could increase the vitality of the host economy by adding to its strength and competitiveness.

While recognising the potential importance of the small and innovative firm to local development, it is also recognised that the evidence to hand suggests that technological change, and notably product development, varies systematically between regions – to the disadvantage of the currently less favoured regions (e.g. Thwaites *et al.* 1981; Oakey *et al.* 1988; Todtling 1990). In general, these studies suggest that the incidence of introducing new, or modified, technology tends to be lower in those regions of any economy that are already economically disadvantaged.

In the absence of any evidence to the contrary, constraints on technological renewal and upgrading would appear to inhibit the long-run development process, but, insofar as local economic development is concerned, it is the economic performance of local firms post-innovation that links the downstream effects of technological advance to the local economy *at least in the first round of effects and in the shorter term*. It is recognised that the subsequent adoption of innovations by imitators or users will also be important, if not more important to *total* economic development (Geroski *et al.* 1993). This diffusion will increase the economic impact in the region if subsequent rounds of effects pass through that region. The research and development stages pre-innovation are obviously also of importance but are not discussed here. However, it has been suggested that industrial innovation is an entrepreneurial act (Thwaites 1977) – as is new firm formation – and it is hypothesised that the return to innovation depends upon the entrepreneurial talents and activities of those in positions of influence within the firm and the entrepreneurial or enterprise culture of the region in which it is located.

The Entrepreneur and the Local Culture of Enterprise

One of the major concerns of this chapter is the influence that the characteristics and decisions of entrepreneurs (represented here by owners/directors) in innovative SMEs may have on the financial performance of the firm. A number of studies (see, for example, Storey *et al.* 1987, 1989; Wynarczyk *et al.* 1993 and Lafuente and Salas 1989) reach broadly similar conclusions, namely that there are differences in the performance of small firms which are related to the characteristics and decisions of the owner.

There appears also to be widespread agreement that entrepreneurial activities both induce and result in innovation, or more generally, economic change. Though as Penrose (1959, p.33) observes, entrepreneurship is: 'a slippery concept...not easy to work into formal analysis because it is so closely associated with the temperament or personal qualities of individuals'. Nevertheless, there have been many theories of entrepreneurship from the 17th century onwards (see, for example, Casson 1990) although it has to be conceded that within modern mainstream economics the concept has generally been neglected (Barreto 1989). Within the economic literature of entrepreneurship it is commonplace to present such economic agents as alert to new opportunities for profit by means of arbitrage, speculation or innovation. The actions of entrepreneurs are seen to play a critical role in overcoming inertia, correcting error and underpinning the systematic discovery of new possibilities. Such actions are necessarily both destructive and constructive, helping to lead the economic system from the old to the new, since their consequences can be either disequilibrating or coordinating. Entrepreneurship is closely associated with economic change, since it induces, responds to and absorbs such change.

The Schumpeterian vision of the entrepreneur highlighted the creative destruction involved in overcoming inertia and the obstacles or barriers to action. Of course, this willingness for change had to be backed up by the effective ability to deliver such change with the provision of resources from a responsive and pliable banking sector. Schumpeter appears to argue that the entrepreneurial function in the world of commodities is equally applicable to the world of ideas. The transformation of productive techniques and process ruptures caused by new innovations and capital re-tooling has its counterpart in the intellectual world. Schumpeter views science as tooled knowledge and associates material transformations with intellectual transformations; the intellectual leadership of the extraordinary scientist bears a striking resemblance to the economic leadership of the entrepreneur (Schumpeter 1934, pp.83–89). There are a number of constraints preventing Schumpeterian intellectual and material innovation from taking place. There is a close attachment to conventional thought and behaviour. As he argued: 'all knowledge and habit once acquired becomes as firmly rooted in ourselves as a railway embankment in the earth' (p.84). We find it difficult to adopt new practices, methods, or views and get locked into established patterns. Material and intellectual innovation are equally risky, requiring the rejection of the old for the new and have, as their consequent, the devaluation or destruction of the old by the application of the new.

Whilst the Schumpeterian entrepreneur struggles to overcome inertia and induce fundamental change, the other key aspects of entrepreneurship highlighting error elimination and the discovery process have been developed within the

Austrian research tradition by Kirzner (1973), amongst others. Apart from emphasising the co-ordinating and deviation-counteracting activity of entrepreneurs, attention has been increasingly focused upon the entrepreneur as the creator or manufacturer of new possibilities. An important development has been the explicit recognition of the entrepreneurial function associated with creative imagination in a world characterised by pervasive uncertainty (see, for example, Loasby 1982; Kirzner 1982). Within such a framework, the entrepreneurial function is carried out by 'individuals' within a societal context. There is a symbiotic relationship between individuals and institutions which denies both the independency of the former and the total determinacy of the latter. Individuals are formed by, and can transform, their institutional setting; likewise, institutions both constrain and enable individual behaviour. All economic agents are social beings and the products of given situational contexts, but action is meaningful or purposive so that change is possible. Decision-making and choices are influenced by the constraints of time, place, resources, endowments and experience. The philosophy of science has characterised such problemsolving as situational logic whereby one explores individual aims, needs or ambitions, and the situational constraints in order to understand the action undertaken (see, for example, Popper 1979, pp.178–190). Preferences, meta-preferences and attitudes to risk and enterprise appear to be heavily influenced by institutional context, conventions, networks and culture. It is in this sense that biography and background matter in the sense of stimulating or constraining certain actions or choices. Entrepreneurship may often involve the transcending of such constraints or inertia through rule-breaking and the creation of new alternative possibilities, whether in the material or intellectual world. The discovery or creative construction of a new reality complements the approach of Schumpeter and endorses the need to overcome barriers to change in order to realise any new possibilities.

The capacity of a region to generate a set of innovative and economically dynamic small firms may be conditioned, therefore, by entrepreneurial talents within the set of potential and existing entrepreneurs in the region and the wider culture of enterprise which surrounds these firms. Storey *et al.* (1987) have shown that major characteristics associated with entrepreneurship vary by region in the UK. For example, the educational standards of people in general in some regions is much below that of other regions and in subjects or disciplines not associated with small firm formation.

Turning from the geographic dimensions to the broader culture of enterprise, Mason (1991), in his review of the literature, suggests that attitude towards, and expectations from, entrepreneurship will influence the social and economic behaviour of individuals and groups in any area and the number, nature and scale of businesses established there. In addition, these 'attitudes' will influence the scale and nature of institutional support for local entrepreneurial activities such as new firm formation and/or innovation. Hisrich and Peters (1992) maintain that some cultures, such as the American culture, '...places a high value on being your own boss, having individual opportunity, being a success, and making money' (p.12) which results in a high level of company formation and innovation in the United States. They also note that within the United States there are local areas in which the entrepreneurial sub-culture is particularly high and supportive of enterprise – giving as examples: Route 128, Silicon Valley, Dallas/Fort Worth and the North Carolina Triangle. At a more micro-scale, Hampden-Turner (1990) argues that: '...corporate cultures act out themes and patterns of the wider culture' (p.12).

Thus the macro or regional culture is inextricably linked to the corporate culture expressed within the local firm. The culture reinforces ideas, feelings and information which are consistent with the beliefs and values of the firm developed from personal characteristics of members of the firm, modulated by the extra-corporate and local culture. It also discourages ideas and possible activities which are inconsistent with the corporate culture. Thus the culture of the firm can support or undermine innovation, professionalisation of activities and procedures or drives for efficiency. These macro and corporate cultures can combine in virtuous or vicious circles of development which will result in benefits or losses to the enterprise or region (Hampden-Turner 1990).

The research reported below is designed to bring together a number issues associated with small-scale technology and entrepreneurship to improve our understanding of the regional development process. This chapter attempts to:

- evaluate the financial performance and entrepreneurial characteristics of innovative small firms in selected manufacturing industries in the UK
- compare and contrast these characteristics of innovative small firms operating in two areas of the country: the 'South-East' region and elsewhere in the UK (i.e. 'Other Regions')
- examine variations in corporate financial performance associated with decisions taken by owners/directors (entrepreneurs) of firms which, in turn, may reflect the varying cultures of enterprises and regions.

The Data Set

The focus of the analysis is on the small-technologically leading manufacturing firm. This type of firm is identifiable on the Science Policy Research Unit's (SPRU) Innovation Database. This database lists 4576 innovations of technological significance to industry and introduced into the UK between the years 1945 and 1983 inclusive. (For a more detailed description of the database see Townsend 1981). The database was estimated to cover the major technological changes taking place in 25 UK industrial sectors which produced approximately 58% of the UK net output in 1975.

For the purposes of this research (i.e., the need to focus on small innovative enterprises) a combination of secondary information sources (e.g. Who Owns Whom, MacCarthy Cards, Jordans, ICC), in conjunction with the SPRU Innovation Database, was used to identify a set of 175 firms which, at the time of innovation, were: independent (directors owned at least 50% of the issued share capital), small (were enterprises, or part of an enterprise, in which total employment was less than 500 employees world-wide) and had introduced products, processes or materials new to the UK in the years 1975 to 1983 inclusive. The selection of this period was designed to allow exploration of the effects of innovation in the period after innovation up to 1989 but would, at the same time, avoid some of the problems associated with assessing events in the very distant past. Secondary and original data and information were collected from firms by postal and interview surveys. In addition, selected information from Companies House was attached to a subset of 51 cases to act as an exploratory study of innovative SME financial performance. It is this last set of information – which includes turnover, assets, profit as well as more qualitative data on directors, shareholdings and changes in holdings, location, etc. – which provides the basis for the analyses of SME financial performances and their relationships to intra-corporate decisions.

The Financial Performance of Innovative SMEs (n=51)

Due to time and resource constraints it has only been possible to assemble and transform for analysis Companies House data on 51 SMEs surviving to 1989 – of which 21 were located in the 'South-East' region, the core economic region, and the remaining 30 firms located in the 'Other Regions' of the UK. For each SME, data were assembled for two periods in their history of operations: at the time of innovation and four years after. In addition, assets and profitability data were collected for the period 1975 to 1987 for a sub-sample of 34 firms which provided data for all the 14 years covered by this study. This choice was made to allow the benefits/costs of innovation to feed through into the accounts. (A list of variables, obtained from Companies House is presented in Appendix 3.1 pp49-50).

One measure of SME financial performance is total asset growth, where assets (i.e. fixed plus current assets) are taken to reflect not only investment but also the 'wealth and size of the firm'. Evidence from our analysis suggests that, in this sample of SMEs, assets have grown quite considerably with two-thirds of firms having assets of over one million pounds each in 1987. Over the 14-year-period covered by our enquiry, assets in monetary terms, and for the average firm, have increased tenfold. This is obviously an over-estimate due to inflation, the effects of which are not easily calculated in an area in which debate continues as to the most appropriate deflators to use, particularly in the small firm sector. However, crude estimates using a number of deflators suggest that real growth in assets for the average firm has taken place. There is, however, some difficulty with the above analysis due to the fact that the innovations took place in different years (i.e. 1975 to 1983) and so the post-innovation experience is unstandardised. In order to overcome this anomaly, data on firms were standardised using percentage changes in financial performance at the time of innovation and four years later. The results are summarised in Table 3.1.

The first row of Table 3.1 illustrates, for all firms, changes in total assets, turnover, retained profits, operating profits and exports in the four year period following innovation. (The formulae and definitions are presented in Appendix 3.1, pp.49–50). The information in Table 3.1 suggests there are considerable inter-firm variations in performance where the mean firm increased total assets by 162% with a median of approximately 90%. The mean change in turnover was not quite so high, but still substantial at 140%, although the median of 89.7% suggests that a few firms are biasing the mean. In the post-innovation period, the average firm seems to have increased exports nearly twice as fast as turnover by increasing its penetration of international markets – where export performance is positively correlated with retained profits.

On the limited evidence available, the surviving and innovative SME would appear to possess a number of high-growth characteristics in terms of financial performance that suggest it is an asset to both national and regional development. However, the pace of regional development is uneven within the UK and it is of interest to discover whether or not the performance of the innovative SME follows

Table 3.1 Univariate results for the financial performance measures of innovative small firms located in the 'south-east' and 'other regions'

	% Change in Turnover		*% Change in Exports*		*% Change in Operating Profits*		*% Change in Retained Profits*		*% Change in Total Assets*	
	Mean	*Median*	*Mean*	*Median*	*Mean*	*Median*	*Mean*	*Median*	*Mean*	*Median*
All firms	140.7	89.7	256.7	120.7	2.6	1.7	-9.7	-0.4	162.6	89.7
South-East firms	105.4	74.8	528.9	255.4	37.8	23.4	27.0	6.2	210.3	89.4
Other Regions firms	168.5	53.2	75.3	69.2	-21.1	0.8	-40.4	-3.6	123.8	90.2
	Z = 0.82		Z = 2.24*		Z = 1.24		Z = 2.022*		Z = 0.88	
	T = 0.85		T = 2.50**		T = 1.50		T = 2.05**		T = 1.4	
	F = 0.50		F = 5.00		F = 2.00		F = 3.00		F = 0.20	

Z = Mann Whitney U Test of the difference between South-East and OtherRegions located firms.

T = Linear regression test of the difference between South-East and OtherRegions located firms.

* = significant at 0.05level of confidence.

** = significant at 95%level

Source: Companies House Data. See appendix 3.1, pp.49–50 for definitions of variables

this pattern or can overcome constraints at the regional level which inhibit other facets of development.

The Regional Dimension

The location of the innovating SMEs shows a tendency for concentration in the south of England (see also Harris 1988). Pearson correlation coefficients presented in Table 3.2 also show that 'South-East' located firms at the time of innovation were larger than their counterparts in the rest of the country in terms of turnover (r=.19) and total assets (r=.18). Using the same criteria and methods of analysis adopted at the national scale, Table 3.1 illustrates the 'regional' dimension to changes in total assets, turnover, retained profits, operating profits and exports in the four-year period following innovation in firms located in the 'South-East' region and 'Other Regions' of the UK.

At the spatial level it would appear that in terms of total assets, operating profits and turnover growth, these are not significantly changed by location. However, there appear to be marked differences between the growth in retained profits and exports for the two groups. For example, the median 'South-East' located firm experienced a positive growth of 6.2% in retained profits over the four years compared with negative growth of -3.6% for the median innovative firm located in the 'Other Regions' over the same period (Table 3.1). Negative growth in retained profit can obviously threaten the viability of the firm in the longer term. The negative growth in retained profits for 'Other Regions' firms can be attributed to the fact that these firms withdrew a relatively higher proportion of operating profit in the form of directors' remuneration than their counterparts in the 'South-East' region during the four-year period under study. Furthermore, the median 'South-East' exporting SME exhibited a positive growth of over 255% in exports compared with the positive growth of nearly 70% for the median 'Other Region' firm (Table 3.1). Thus, while growth of turnover is not significantly different between firms located in the 'South-East' or 'Other Regions', 'South-East' located firms, particularly those with higher total assets, seem to focus on exports with greater success than their counterparts in the rest of the UK. At the time of innovation, and for firms in both regions, exports accounted for approximately 31% of sales. Four years later, for the 'Other Regions' firms, the proportion of sales to overseas markets remained unchanged while firms in the 'South-East' had increased exports to nearly 50% of output. The evidence of high involvement with and growth of exports in the innovative SME sector would seem to allow them to be classified (on one characteristic at least) as part of the fast-growth firm sector (Storey *et al.* 1987). In particular, the 'South-East' region, in comparison with the 'Other Regions', appears to be benefiting from this activity (r=.52) (Tables 3.2a and 3.2b).

Tables 3.2 Pearson Correlation Matrix (all companies)

a. At the time of innovation

	REG	TA1	F1	NS1	EQ1	SD1	RD1	ND1	S1	OP1	RP1
TA1	.18										
F1	-.31	-.11									
NS1	.04	.40	-.05								
EQ1	-.04	-.33	.19	-.56							
SD1	.31	.15	-.11	.25	.15						
RD1	-.25	-.06	.57	-.10	.21	.21					
ND1	.05	.40	.16	.69	-.23	.27	.25				
S1	.19	.90	-.12	.51	-.36	.16	-.04	.32			
OP1	.24	.22	-.27	-.05	.03	.07	-.25	-.11	-.01		
RP1	.14	-.07	-.21	-.20	.14	-.03	-.21	-.22	-.27	.93	
EX1	.03	.76	.18	.43	-.70	.42	.20	.21	.70	.42	.30

Source: Companies House Data. See Appendix 3.1, pp.49–50 for definition of variables

b. Four years after the innovation

	REG	TA4	F4	NS4	EQ4	SD4	RD4	ND4	EX4	CRP	COP
TA4	.25										
F4	-.23	-.14									
NS4	.17	.34	-.24								
EQ4	-.19	-.05	.10	-.44							
SD4	.32	.40	-.18	.31	.08						
RD4	-.31	-.14	.65	-.21	.30	-.07					
ND4	.23	.41	-.02	.56	-.29	.30	.03				
EX4	.41	.83	-.34	.61	-.51	.58	-.36	.48			
CRP	.30	.14	.10	-.08	.05	.42	.13	.22	.45		
COP	.20	.06	.06	-.12	.14	.01	-.08	.10	.09	.92	
CEX	.52	-.02	-.25	.22	-.49	.39	.20	.19	.02	.71	.23

Source: Companies House Data. See Appendix 3.1, pp.49–50 for definition of variables

Entreneurship and the Regional Culture of Enterprise

It would be foolish to claim that the research reported here fully explores the cultural issues in economic development, either within the enterprise or in the wider (regional) operating environment. Instead it explores associations between the financial performance of the innovative SME against a number of selected characteristics or decisions made within the firm, which might be reflective of its entrepreneurial culture and, in aggregate, the enterprise culture of the region in which it is located.

The literature on entrepreneurs repeatedly stresses 'independence' as the primary motivating force of individuals setting up and retaining control of small firms. While the notion of independence and autonomy can be an admirable trait, it has also been shown that the desire to retain control can inhibit the development of the firm – particularly when this results in undercapitalisation. It might be hypothesized that those firms which severely restrict the number of shares issued and limit their sale to a few individuals, particularly family members, are less entrepreneurial and less growth oriented than those firms which take a more outward looking stance to introduce new share ownership and nonfamily members to share ownership. The results of our analysis show that the innovative SMEs doubled the total number of shares issued between the time of innovation and four years later. At the sub-national scale, 'South-East' firms issued significantly more shares than those firms located in other parts of the UK both before and after innovation.

Some watering-down of ownership is evident across the UK but it appears more evident in the 'South-East' region than in the remainder of the country. This evidence may suggest that first, directors and owners of 'South-East' located firms are more willing to appoint new shareholding directors post-innovation and second, to turn to the equity market and in order to raise funds to support entrepreneurial activities than are their counterparts in the rest of the country (Mason and Harrison 1992). On the other hand, it could reflect regional variations in the availability of capital for small and innovative ventures (Mason and Harrison 1991).

An important characteristic which may differentiate between firms in their approach to growth is not only the different background of the entrepreneurs but also the type of individuals they are willing to bring into senior positions once the firm has become involved in innovative activities. As Grieve-Smith and Fleck (1987) correctly point out, managerial appointments are central to the growth of small firms – particularly innovative small firms (Thwaites and Wynarczyk 1994). The ability and willingness to innovate is only a necessary, and not a sufficient, condition for growth. It may be hypothesized that innovation leads to further growth. Growth can introduce crises which impose pressures on existing management for additional managerial resources in order to cope with the new situations. Only if additional, suitably qualified and motivated managerial talents become available, from either internal and/or external sources, will these crises be satisfactorily overcome and further growth facilitated (Wynarczyk *et al.* 1993).

By law, small independent limited companies in the UK are required to set up and trade with a minimum of two shareholding directors. The majority of small firms in the UK are set up with this minimum required number of directors – usually husband and wife teams (husband holds 99% and wife 1% of the issued share capital), or two individuals (e.g. father and son, two colleagues). For the typical small firm, the ownership and shareholding structure remains unchanged throughout its life span. However, an examination of data reveals that only around 20% of the innovative small firms in our sample had only two directors at the time of innovation. Moreover, they tended to appoint new and working directors to the board on an increasing scale in the post-innovation period. This is particularly noticeable in the 'South-East' where, four years after innovation, over 60% of the innovative small firms located in the 'South-East' region had more than five directors, compared with just over 43% of their counterparts in the 'Other Regions'. Moreover, our evidence suggests (Tables 3.3 and 3.4) that, for our sample of firms, husband and wife teams of directors are more prevalent in the 'Other Regions', while technical directors are more commonly found in the 'South-East'. Such differences continue over time and through the innovation to the post-innovation period, even when the number of directors could be expected to increase as firms expand output. 'South-East' firms employ a significantly greater number of technical and scientific directors (r=.32) (Table 3.2, p.41) and a significantly lower number of female (r=-.23) and related (wives/siblings) (r=-.31) directors than firms located elsewhere in the UK. 'South-East' located firms with the greatest number of non-shareholding directors, and those with the greatest number of

Tables 3.3 Pearson Correlation Matrix ('south-east' firms)

a. At the time of innovation

	TA1	F1	NS1	EQ1	SD1	RD1	ND1	S1	OP1	RP1
F1	-.22									
NS1	.33	.07								
EQ1	-.17	.17	-.75							
SD1	.23	.23	.43	-.03						
RD1	-.18	.69	.04	.18	.14					
ND1	.29	.22	.76	-.44	.41	.22				
S1	.95	-.14	.38	-.15	.09	-.15	.31			
OP1	.83	-.19	.37	-.23	.30	-.14	.27	.50		
RP1	.78	-.17	.26	-.13	.29	-.03	.24	.84	.83	
EX1	.47	.49	.95	-.67	.78	.48	.89	.43	.32	.36

Source: Companies House Data. See Appendix 3.1, pp.49–50 for definition of variables

b. Four years after the innovation

	TA4	F4	NS4	EQ4	SD4	RD4	ND4	EX4	COP	CRP	CEX
F4	-.14										
NS4	.33	-.22									
EQ4	.16	.09	-.46								
SD4	.54	.08	.31	.21							
RD4	-.08	.88	-.10	-.03	-.02						
ND4	.32	.17	.51	-.32	.28	.20					
EX4	.80	.21	.88	-.20	.67	.34	.74				
COP	-.17	.04	.12	-.24	.14	-.10	-.05	-.01			
CRP	-.20	-.17	.03	-.62	.04	-.22	-.02	-.21	.38		
CEX	-.33	.14	.33	-.62	.50	.26	-.02	-.22	.70	.78	
CTA	-.20	.02	.11	-.15	.21	.10	.11	.17	.63	.54	.78

Source: Companies House Data. See Appendix 3.1, pp.49–50 for definition of variables

technical and scientific directors, also have the highest exports (r=.88 and r=.67 respectively; Tables 3.3). This is also, to some degree, the case for firms in the rest of the country, where the higher the number of technical and scientific directors, the higher the exports (r=.26; Table 3.4), in comparison with those firms with strong family connections (r=.14).

Tables 3.4 Pearson Correlation Matrix ('other regions' firms)

a. At the time of innovation

	TA1	F1	NS1	EQ1	SD1	RD1	ND1	S1	OP1	RP1
F1	.15									
NS1	.50	-.11								
EQ1	-.52	.20	-.38							
SD1	-.12	-.19	.03	.17						
RD1	.29	.47	-.04	.23	.12					
ND1	.58	.17	-.11	-.10	.17	.29				
S1	.77	.06	.68	-.53	.10	.18	.36			
OP1	-.24	-.20	.24	-.24	.18	-.21	-.25	.58		
RP1	-.32	-.20	-.38	-.26	.15	-.21	-.29	-.64	.99	
EX1	.98	.20	-.22	-.77	.34	.14	-.30	.96	.58	.59

Source: Companies House Data. See Appendix 3.1, pp.49–50 for definition of variables

b. Four years after the innovation

	TA4	F4	NS4	EQ4	SD4	RD4	ND4	EX4	COP	CNP	CEX
F4	-.04										
NS4	.33	-.20									
EQ4	-.26	.01	-.39								
SD4	-.04	-.31	-.16	.11							
RD4	-.17	.49	-.06	.11	-.02						
ND4	.56	-.04	.56	-.22	.22	.00					
EX4	.92	.15	.10	-.49	.26	.14	.20				
COP	.14	-.16	-.21	-.27	.26	.06	.10	-.32			
CNP	.20	.10	.50	-.29	.19	.12	.54	.58	.97		
CEX	.09	.10	.51	.20	.24	.12	.55	.09	-.59	-.29	
CTA	.22	-.54	-.08	.12	.41	-.43	-.15	-.23	-.10	-.25	.4

Source: Companies House Data. See Appendix 3.1, pp.49–50 for definition of variables

In the case of these innovative SMEs, there is also a tendency over time for them to take on professional directors with managerial and finance expertise, with a proportion of them also becoming shareholders. The multiple regression results presented in Table 3.5 reveals that new directors appointed post-innovation are more likely to be those from professional backgrounds. Furthermore, our data also suggests that those firms with a combination of technical and professional directors appear to have the highest growth in exports, turnover and retained profits than other firms in our sample with different ownership structures. These firms are more likely to be located in the 'South-East' regions.

Table 3.5 Multiple regression results

Variables		*T Statistics*	
	DEP = RD4	*DEP = SD4*	*DEP = DPT4*
CEX	1.2	2.1***	2.8***
CRP	-.59	2.5***	2.9***
F4	3.5***	1.4	1.3
SD4	-1.3	–	2.5***
RD4	–	-1.5	-1.4
PD4	-.63	1.6	–
ND4	.8	1.8	2.1***
F =	1.7	6.4	8.7
$\overline{R}^2$	.39	.65	.67

Source: Companies House Data. See Appendix 3.1, pp.49–50 for definition of variables
*** = Significant at one per cent level

Another aspect of behaviour/culture which may affect the potential for, and actual level of, growth in the firm is the decision of the directors to retain high levels of earned income in the firm in order to support expansion, and/or risky or long-term ventures – including technological change and product development. One means of building up funds for such ventures is to build up retained capital. Owners/Directors can contribute significantly to this process through their decisions on what they pay themselves, and/or through personal loans to the firm.

As we saw earlier, there was a significant difference between firms in the growth of retained profits – to the advantage of those located in the 'South-East' region (Table 3.1, p.39). An examination of the data shows that directors of innovative SMEs located in the 'South-East' region averaged £11,182 at the time of innovation, compared with £9483 in 'Other Regions'. Post-innovation, the rate of increase in directors remuneration was greater in the 'Other Regions' (111% over four years) to that in the 'South-East' (101%). While these are not significantly different sums, the effect is to leave 'South-East' firms in a stronger financial position, measured in terms of growth of retained profits and assets, than their counterparts in other parts of the country.

Summary and Conclusions

The research has attempted to contribute to the debate surrounding regional economic development and, in particular, the role of the innovative SME in this process. The research concentrates on first-round effects of innovation on SME development, which indicates that the average survivor has experienced considerable growth in assets, return on assets, retained profits and exports to a greater degree than those firms identified by Storey *et al.* (1987) as 'high growth' firms or more run-of-the-mill firms. On the limited evidence available here, the surviving and innovative SME appears to be part of the set of fast-growth firms and, as such, would seem to warrant the attention it receives from policy makers and academics alike.

At the regional level (i.e. 'South-East' v. 'Other Regions'), the evidence suggests that significant innovations are more likely to be introduced into the 'South-East' region than elsewhere in the country. The research also showed that retained profits and exports grow more strongly in firms located in the 'South-East' region than in those located in 'Other Regions' of the UK. On these measures, the innovative SME does not seem able to shake off the heritage of its environment in other parts of the UK to out-perform, on average, its counterparts in the core and 'South-East' region.

The relationship between financial performance and characteristics or decisions of the owners/directors, where the latter is seen as indicative of the corporate entrepreneurial culture, was tentatively explored at the regional level. While the general perception of the owner of a small firm is one who values highly his independence, the evidence reported here suggests that owners of innovative SMEs are willing to issue relatively large numbers of additional shares, which, in aggregate, were seen to double in the four years after innovation. At the same time, the owners were willing to let their proportion of equity-holding reduce, with some giving up overall control. The issue of shares and loss of total control was most pronounced in the 'South-East'. High share retention by owners was negatively correlated with export performance.

The directors of companies located elsewhere in the UK were more likely to be family-run, which was significantly different from those in the 'South-East' – which employed more technical and scientific directors. These professional directors were closely associated with exports and profitability growth, in contrast to the family-run companies. Furthermore, the results of this study also suggest that, over time and post-innovation, those firms with technical directors are more likely to recruit directors with managerial and finance expertise from outside to complement the technical expertise on which their firms were based. This type of innovative small firm, which is located mainly in the 'South-East' region, seems to be more successful, and grows more strongly in terms of exports, turnover and profit, than their counterparts with different ownership structures.

Retained profits are a potential source of investment in expansion or risky ventures such as product development. We have already noted that retained profits grew most strongly in the 'South-East'. We also found that directors in 'South-East' firms, and particularly those companies engaging technical, scientific and nonfamily directors, were willing to curb their own income from the firm to invest in its future. While explanations still remain elusive, there do appear to be significant spatial differences in the ways in which innovative SMEs are financed and operated by their owners.

In conclusion, it is hoped that this research allows a small step forward in our understanding of the operation of innovative SMEs and their links to local economic development. The policy-maker would seem justified in supporting such firms but there are still sectoral, spatial, entrepreneurial and local culture of enterprise factors which appear to influence economic outcomes and are worthy of further research.

References

Barreto, H. (1989) *The Entrepreneur in Microeconomic Theory.* London: Routledge.

Birch, D. (1979) *The Job Generation Process.* Cambridge, Mass: MIT Programme on Neighbourhood and Regional Change.

Casson, M. (1990) *Entrepreneurship.* Aldershot: Edward Elgar.

Elliott-White, M. (1991) Unpublished PhD Thesis, Humberside College of Higher Education.

Geroski, P., Machin, S., and Vanreenen, J. (1993) 'The profitability of innovating firms.' *Rand Journal of Economics 24,* 2, 198–211.

Gray, C. (1992) 'Growth orientation and the small firm.' In K. Caley, E. Chell, F. Chittendon and C. Mason (eds) *Small Enterprise Development.* London: Paul Chapman.

Grieve-Smith, J. and Fleck, V. (1987) 'Business strategies in small high technology companies.' *Long Range Planning 20,* 2, 61–8.

Hampden-Turner, C. (1990) *Corporate Culture.* London: Hutchinson.

Harris, R.D. (1988) 'Technological change and regional development in the UK: evidence from the SPRU database on innovations.' *Regional Studies 22,* 5, 361–374.

Hisrich, R. and Peters, M. (1992) *Entrepreneurship: Starting, Developing and Managing a New Enterprise,* second edition. Boston: Irwin.

Kirzner, I.M. (1973) *Competition and Entrepreneurship.* Chicago: Chicago University Press.

Kirzner, I.M. (1982) 'Uncertainty, discovery and human action.' In I.M. Kirzner (ed) *Method, Process, and Austrian Economics.* Lexington: Lexington Books.

Lafuente, A. and Salas, V. (1989) 'Types of entrepreneurs and firms.' *Strategic Management Journal 24,* 689–713.

Loasby, B.J. (1982) 'Economics of dispersed and incomplete information.' In I.M. Kirzner (ed) *Method, Process, and Austrian Economics.* Lexington: Lexington Books.

Malecki, E. (1991) *Technology and Economic Development.* Harlow: Longman.

Mason, C. (1991) 'Spatial variations in enterprise: the geography of new firm formation.' In R. Burrows (ed) *Deciphering the Enterprise Culture.* London: Routledge.

Mason, C. and Harrison, R. (1991) 'Venture capital, the equity gap and the north-south divide in the UK.' In M. Green (ed) *Venture Capital: International Comparisons.* London: Routledge.

Mason, C. and Harrison, R. (1992) 'A strategy for closing the small firms finance gap.' In K. Caley, E. Chell, F. Chittendon and C. Mason. (eds) *Deciphering the Enterprise Culture.* London: Routledge.

Oakey, R.P., Rothwell, R. and Cooper, S. (1988) *Management of Innovation in High Technology Small Firms.* London: Pinter.

Pavitt, K., Robson, M. and Townsend, J. (1987) 'The size distribution of innovating firms in the UK: 1945–1983.' *Journal of Industrial Economics 35*, 3, 297–316.

Popper, K.R. (1979) *Objective Knowledge.* Oxford: Clarendon Press.

Penrose, E.T. (1959) *Theory of the Growth of Firm.* Oxford: Basil Blackwell.

Rosenburg, N. (1992) 'Science and technology in the twentieth century.' In G. Dosi, R. Gianetti and P. Tominelli. *Technology and Enterprise in a Historical Perspective.* Oxford: Clarendon Press.

Rothwell, R. (1986) 'The role of small firms in technological innovation.' In J. Curran, J. Stanworth and D. Watkins (eds) *The Survival of the Small Firm 2.* Aldershot: Gower.

Schumpeter, J.A. (1934) *The Theory of Economic Development.* Harvard: Harvard University Press.

Storey, D.J., Keasey, K., Watson, R. and Wynarczyk, P. (1987) *The Performance of Small Firms.* London: Croom Helm.

Storey, D.J., Watson, R. and Wynarczyk, P. (1989) *Fast Growth Small Businesses.* Research Paper No.67, Department of Employment, Sheffield.

Thwaites, A.T. (1977) *The Industrial Entrepreneur: a Definitional Problem.* Discussion Paper No. 4, CURDS, University of Newcastle upon Tyne.

Thwaites, A.T., Oakey, R.P. and Nash, P.A. (1981) *Industrial Innovation and Regional Development.* Final Report to Department of the Environment, CURDS, University of Newcastle upon Tyne.

Thwaites, A.T. and Wynarczyk, P. (1996) 'The performance of innovative small firms in the "South-East" and elsewhere in the UK.' *Regional Studies 30*, 2, 135–149.

Todtling, F. (1990) 'Regional differences and determinants of entrepreneurial innovation: empirical results of an Austrian case study.' In E. Cicioti, N. Alderman and A. Thwaites (eds) *Technological Change in a Spatial Context.* Heidelberg: Springer-Verlag.

Townsend, J. (1981) *Innovations in Britain since 1945.* Occasional Paper No.16, SPRU, University of Sussex, Brighton.

Wynarczyk, P., Watson, R., Storey, D.J., Short, H. and Keasey, K. (1993) *Managerial Labour Markets in the Small Firm Sector.* London and New York: Routledge.

Appendix 3.1

List of Variables

AGE: age of firms in years;
Y1: At the time of innovation;
Y4: Four years after the innovation;
SE: 'South-East' located firms (coded as 2);
OT: 'Other Regions' located firms (coded as 1);

Financial Data

TA1: Total Assets at year 1 (representing the size of the firm at the time of innovation);
TA4: Total Assets at year 4 (representing the size of the firm four years after the innovation);
CTA: % change in total assets ((TA4 -TA1/TA1)*100);
EX1: Total Exports at the time of innovation;
EX4: Total Exports four years after the innovation;
CEX: % change in exports ((EX4-EX1)/EX1)*100);
S1: Turnover sales at the time of innovation;
S4: Turnover sales four years after the innovation;
CS: % change in turnover ((S4-S1)/S1)*100);
RP1: Retained Profits at the time of innovation;
RP4: Retained Profits four years after the innovation;
CRP: % change in retained profits ((RP4-RP1)/TA1)*100);
OP1: Operating Profits at the time of innovation;
OP4: Operating Profits four years after the innovation;
COP: % change in operating profits ((OP4-OP1)/OP1)*100);

Non-Financial Data

EQ1: Equity held by the directors and the members of their families at the time of innovation (taken as the % of the total shares issued and fully paid);
EQ4: % Equity held by the directors four years after the innovation;
ND1: Total no. of directors at the time of innovation;
ND4: Total no. of directors four years after the innovation;
RD1: Total no. of related directors at the time of innovation;
RD4: Total no. of related directors four years after the innovation;
F1: Female directors at the time of innovation;
F4: Female directors four years after the innovation;
NS1: Total no. of non-shareholding directors at the time of innovation;
NS4: Total no. of non-shareholding directors four years after the innovation;

SD1: Total no. of scientific and highly technical directors at the time of innovation;
SD4: Total no. of scientific and highly technical directors four years after the innovation;
PD1: Total no. of professional directors at the time of innovation;
PD4: Total no. of professional directors four years after the innovation.
PDT4: SD4 + PD4

CHAPTER 4

Technopolitan Spaces in the Greater Paris Region and the International Restructuring of Firms

Jeanine Cohen

As a work-place, the metropolitan space of Greater Paris is shaped by socio-professional and functional differences (Cohen 1991; 1993). Among the main factors that differentiate its geographical sectors in this respect, one can see the local job commitment to Research and Development and the high levels of executives and technicians (Figure 4.1). Widely spread in the southern and western suburbs and outer metropolitan areas, this phenomenon has several focuses (Figure 4.2), one of which, the Saclay Plateau, has acted as a magnet for new industrial settlements since the 1960s, and has developed into a science park with a scientific university and public research laboratories. After very noticeable development during the 1970s and 1980s, the firm's dynamics are slowing down there, as well as in Greater Paris as a whole, though the geographical sector is still one of the healthiest of the capital region. Even if it appears as an 'excellence pole', it is not disconnected from the French and Parisian evolutions. The transformations of its activities are part of the broader restructuring that affects Greater Paris as a whole, and that results not only in a modernisation of the productive system, but also in heavy job-losses.

An attempt to address the questions raised by this restructuring at the region's level has been done by the planners: at the end of the 1980s, the Government asked the planning services of the state (Direction Régionale de l'Equipement d'Ile-de-France), of the region (Institut d'Aménagement et d'Urbanisme de la Region Ile-de-France) and of the City of Paris (Atelier Parisien d'Urbanisme) to undertake surveys and prepare, in co-operation, a joint project for the new planning and urbanism guidelines (Schéma Directeur d'Aménagement et d'Urbanisme). As to the Saclay Plateau area, that appeared as the heart of the francilian technopolitan spaces, the Government asked the fourteen *Communes* authorities of this area to present their own joint project. Consequently, from 1992, the shaping of the area during the next twenty-five years has, and will have, to comply with this local Schéma Directeur d'Aménagement et d'Urbanisme that matches the regional one, and that includes new public initiatives such as the building of some infrastructure. As a result, the regional future of the Ile-de-France Region seems to be reasonably prepared, and the Saclay Plateau's too.

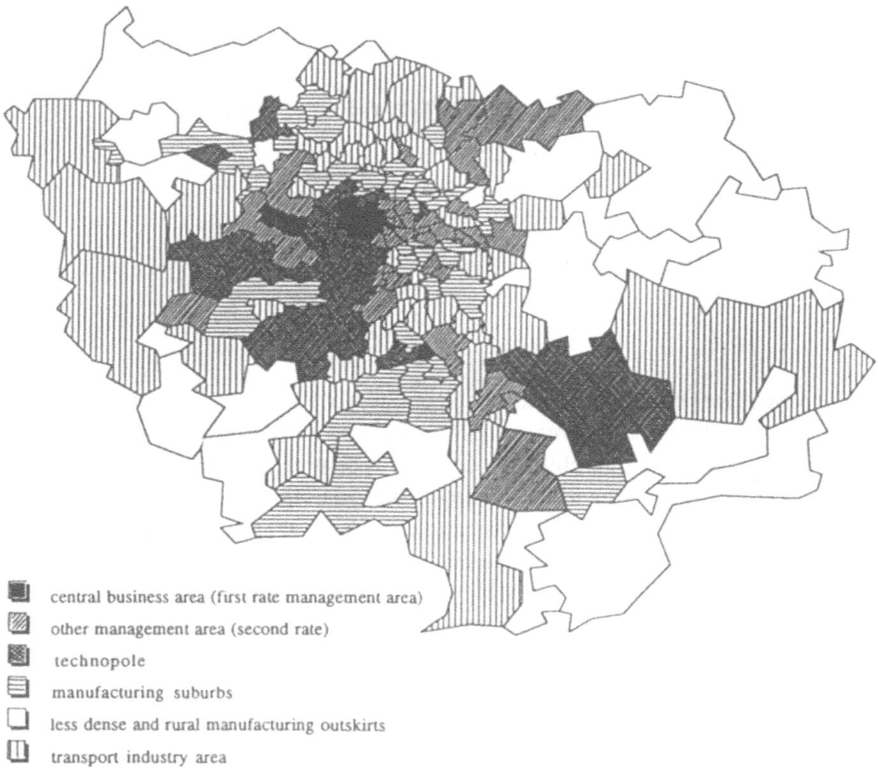

Source: elaborated from the 'Enquête sur la Structure des Emplois' data. (After Cohen, J. (1992) Annales de la Recherche urbaine no50, p.45)

Figure 4.1 Jobs structures in the geographical sectors Ile-de-France Region, 1988

But, the regional and local futures do not only rely on the planners' decisions. Despite the good co-operation between the technical planning services of the different tiers of government acting inside the Ile-de-France regional area, and between the fourteen *Mairies* of the Plateau de Saclay *Communes*, there is a challenging situation. The Saclay Plateau evolution, as well as the future of the Ile-de-France region in its entirety, are strongly determined by Greater Paris' general evolution. Greater Paris at present includes some communes from neighbouring regions, such as Champagne-Ardennes, Haute-Normandie and Centre – if we include the 1990 limits of the industrial and urban populating area (zone de peuplement industriel et urbain, ZPIU) (Damette, Scheibling *et al.* 1992). This area is probably too large (maybe still too dense), its inner differences are too important, and the political options of the local and regional tiers of government are too diverse to allow close co-operation on an easy and long-term basis.

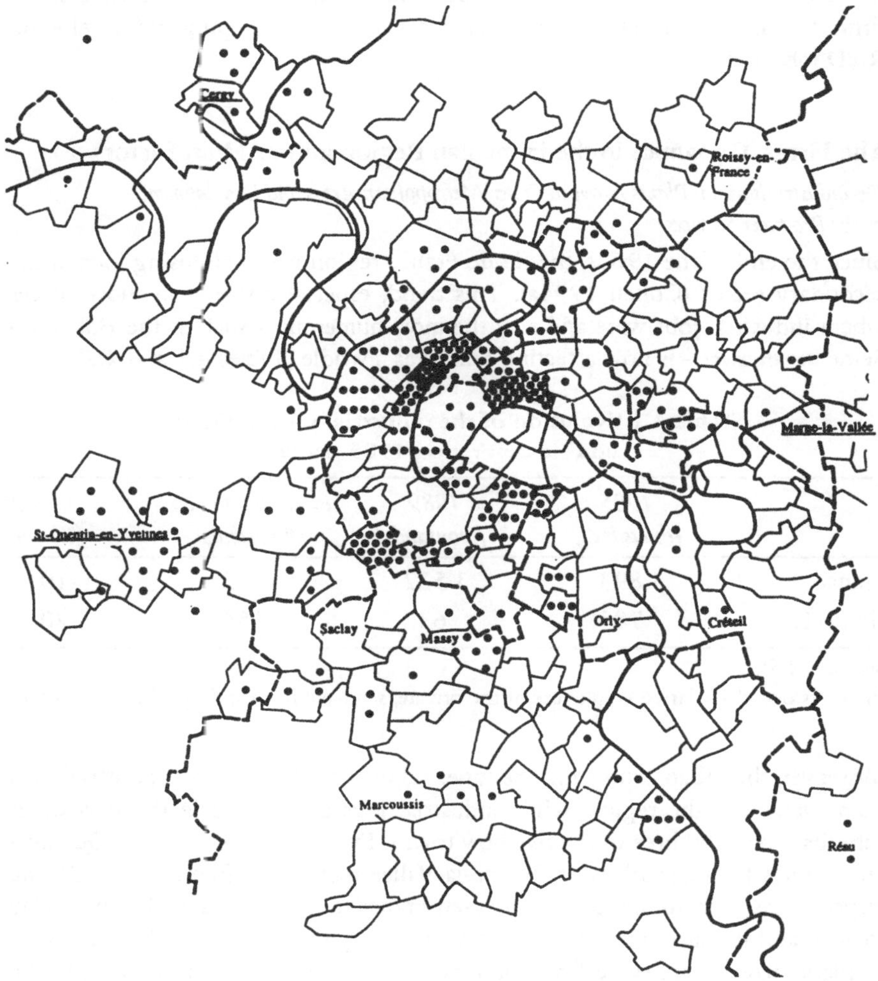

Source: elaborated from the 'Enquête sur la Struture des Emplois' data (After Cohen, J. (1987) in Actes du colloque international 'Villes et technopoles', CNRS-CIEU, Toulouse F. 23–25 September, Jalabert, G. and Thouzellier, C. (eds) P.U.M.)

Figure 4.2 Francilian high-technology premises staff over 100 people, 1983 (staff over 100 people, engineers and production executives rate over the regional average, 7%)

The firms' restructuring has its own dynamics and the supply of local infrastructure is not the only factor that acts upon it. I shall try to discover the interrelations between activities and jobs evolutions, on the one hand, and the firms' strategies on the other. To what extent do the present acquisitions, deregulations and privatisations modify the role and place of R&D in Greater Paris?

Second, I shall explain why and how, in this framework, the Saclay Plateau developed as a technopole. Given the other factors reshaping the presence of the firms, I shall try to assess whether local actors can help in keeping and developing R&D jobs and activities.

The Firms' Dynamics in the Francilian Region and its Main Factors

De-industrialisation: Plants Move to Outer Metropolitan Areas and Development of the Producer Services

Since the end of the 1960s, the Ile-de-France region has been losing jobs in the secondary sector (Cohen 1994a). This trend, especially strong in those places where industrial jobs were and stay the most numerous – such as the Hauts-de-Seine *departement* – is now affecting France as a whole (Tables 4.1 and 4.2).

Table 4.1 Evolution of the industrial jobs in France and Ile-de-France, 1985–1989

	1985 (thousands)	*1989 (thousands)*	*evolution 85–89 (thousands)*	*(%)*
France:	3 893	3 522	-372	-10
Ile-de-Fr.:	785	632	-154	-20

Source: ESE

Premises over 19 salaried employees of the private, semi-public or industrial public sector

Moreover, the urban centre and the inner suburbs are losing more industrial jobs than average in the region, while industrial jobs are created in the new outer suburbs – namely three of the five new towns: Evry to the south, Saint-Quentin-en-Yvelines to the south-west, Marne-la-Vallee to the east (see Figure 4.3); the airport areas: Orly to the south, Roissy-en-France to the north; and also the Saclay Plateau and its surroundings between Evry and Saint-Quentin-en-Yvelines. A lot of plants have disappeared from the inner suburbs, some of them moving to the outer suburbs. The eviction of secondary-industry jobs from the urban centres is not new and one can note, at present, not only a renewal of the garment industry in the north-eastern part of the old Paris centre but an increase in industrial jobs in the very new business centre, Courbevoie, which is part of La Defense or close to this centre. The three neighbouring departments of Paris (Hauts-de-Seine, Seine Saint-Denis and Val-de-Marne) had long been the realm of industrial plants and warehouses. Now, be it by the closure of a number of those premises or by the transformation of their job structures, they are changing and plants and warehouses are to be found more and more in the outer suburbs (Beckouche and Cohen 1992).

Table 4.2 Industrial jobs in the departments of Ile-de-France

a. Weights: Evolution 1985–1989

Department	*1989*		*evolution 1985–89*	
	(thousands)	*(%)*	*(thousands)*	*(%)*
92-Hauts-de-Seine	167	26.5	-58	-26
75-Paris	111	17.6	-52	-32
78-Yvelines	85	13.4	-6	-7
93-Seine-Saint-Denis	70	11.2	-18	-20
91-Essonne	56	8.9	-2	-3
77-Seine-et-Marne	56	8.9	-4	-7
94-Val-de-Marne	46	7.2	-10	-18
95-Val-d'Oise	40	6.3	-3	-8
Ile-de-France Region:	**631**	**100**	**-153**	**-20**

b. Structures (by functions and qualifications), 1989

	departments							
	92	*75*	*78*	*93*	*91*	*77*	*94*	*95*
% jobs by functions:								
Management:	35	44	25	22	24	15	24	21
Marketing:	20	24	7	9	11	6	13	9
Conception:	16	3	15	8	12	7	7	10
Service:	2	4	3	3	2	2	3	2
Warehousing-transport:	3	3	5	7	6	8	8	8
Manufacturing:	24	23	46	50	45	61	45	49
	100	**100**	**100**	**100**	**100**	**100**	**100**	**100**
% jobs by qualifications:								
Executives-engineers:	32	30	23	15	22	11	19	16
Technicians:	34	27	25	24	29	19	23	25
Clerks:	14	21	10	11	11	10	13	11
Skilled workers:	15	17	30	35	28	38	31	35
Unskilled workers:	5	5	12	15	10	22	14	13
	100	**100**	**100**	**100**	**100**	**100**	**100**	**100**

Table 4.2 Industrial jobs in the departments of Ile-de-France (continued)

c. Structures: evolution 1985–1989 (% points number)

	departments							
	92	*75*	*78*	*93*	*91*	*77*	*94*	*95*
jobs by functions:								
Management:	+4	+1	+3	+1	+1	–	–	–
Marketing:	+7	+4	+1	+1	+2	+1	+3	+1
Conception:	+1	-2	+2	–	-1	-1	–	+1
Service:	–	–	–	–	–	–	–	-1
Warehousing-transport:	-1	–	-1	–	–	–	–	–
Manufacturing:	-10	-3	-5	-3	-1	–	-4	-1
jobs by qualifications:								
Executives-engineers:	+9	+3	+7	+3	+4	+2	+5	+3
Technicians:	+3	-2	+3	+1	+2	-1	-1	+1
Clerks:	-5	–	-2	–	-2	+1	-2	-1
Skilled workers:	-5	–	-2	–	-2	+1	-2	-1
Unskilled workers:	-5	–	-5	-2	-1	-1	–	-1

Source: ESE
Premises over 19 salaried employees of the private, semi-public or industrial public sector.
92=Hauts-de-Seine, 75=Paris, 78=Yvelines, 93=Seine-Saint-Denis, 91=Essonne, 77=Seine-et-Marne, 94=Val-de-Marne, 95=Val-d'Oise.

While secondary industry is losing more and more jobs, the producer-services are gaining new ones (Table 4.3).

Table 4.3 Compared evolutions of secondary industry and producer-services jobs in the Ile-de-France region, 1977–1991 (thousands)

	1977–1985	*1985–1989*	*1989–1990*	*1990–1991*
Secondary industry	-1300	-130	+35	-166
Producers services	+180	+400	+60	+6

Source: ASSEDIC
Note: the observation ends in 1991 because of the changing of activities code after that year. Between 1992 and 1993, there was a slight decrease in the total jobs of the new code's producer-services

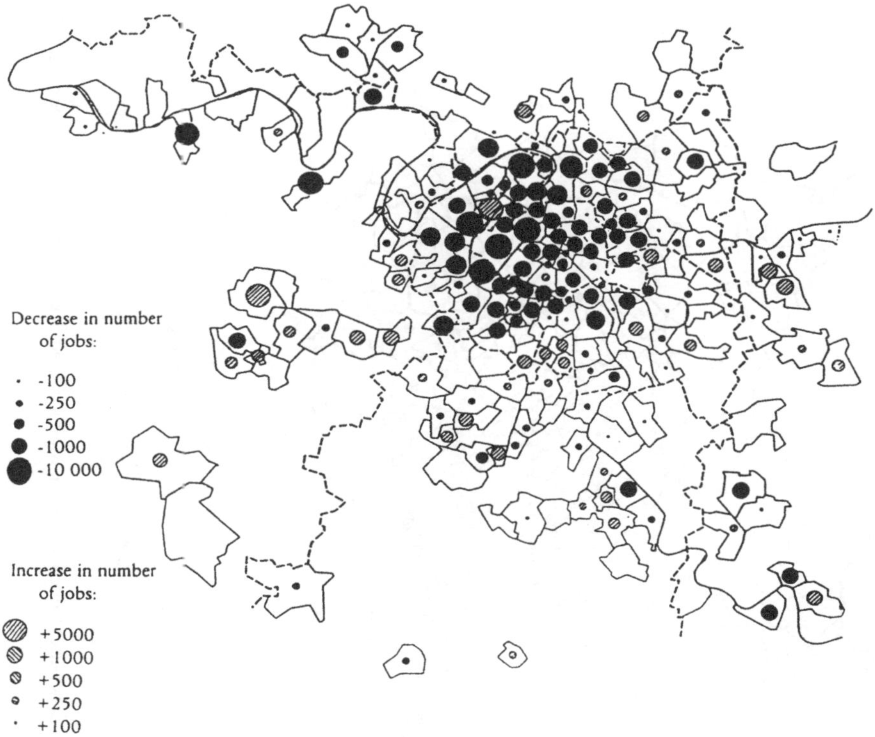

Source: elaborated from the 'Enquête sur la Struture des Emplois' data. (After Cohen, J. (1994) in Insee – Regards sur I'Ile-de-France no25, p.16)

Figure 4.3 Industrial jobs change, March 1985–December 1989, in the Paris agglomeration and neighbouring communes (over 450 jobs in premises over 19 salaried, civil servants excluded)

It was mainly Paris that became more and more dedicated to this type of industry. They are still concentrated in the old right bank centre: 1st, 2nd, 8th, 9th and 10th *arrondissements*. In the 1980s a number of new ones were created in the other arrondissements, along the River Seine at La Defense and its surroundings, as well as in the neighbour communes of Paris and in, or near, the new outer employment areas such as the new towns or the Saclay Plateau, at Les Ulis (Figure 4.4).

The figures in Table 4.3 (p.56) show that one cannot say that services are replacing the secondary-industry jobs. To the opposite, one can guess that producer-services development dynamised the secondary-industry jobs, but those services are now facing a slowing-down that is dramatically amplified in the industry.

Source: elaborated from the 'Enquête sur la Structure des Emplois' data (After Cohen, J. (1993), Métropoles en déséquilibre?, Paris, Economica, p.215)

***Figure 4.4** Producer-services jobs in the Ile-de-France region (1977–1985)*

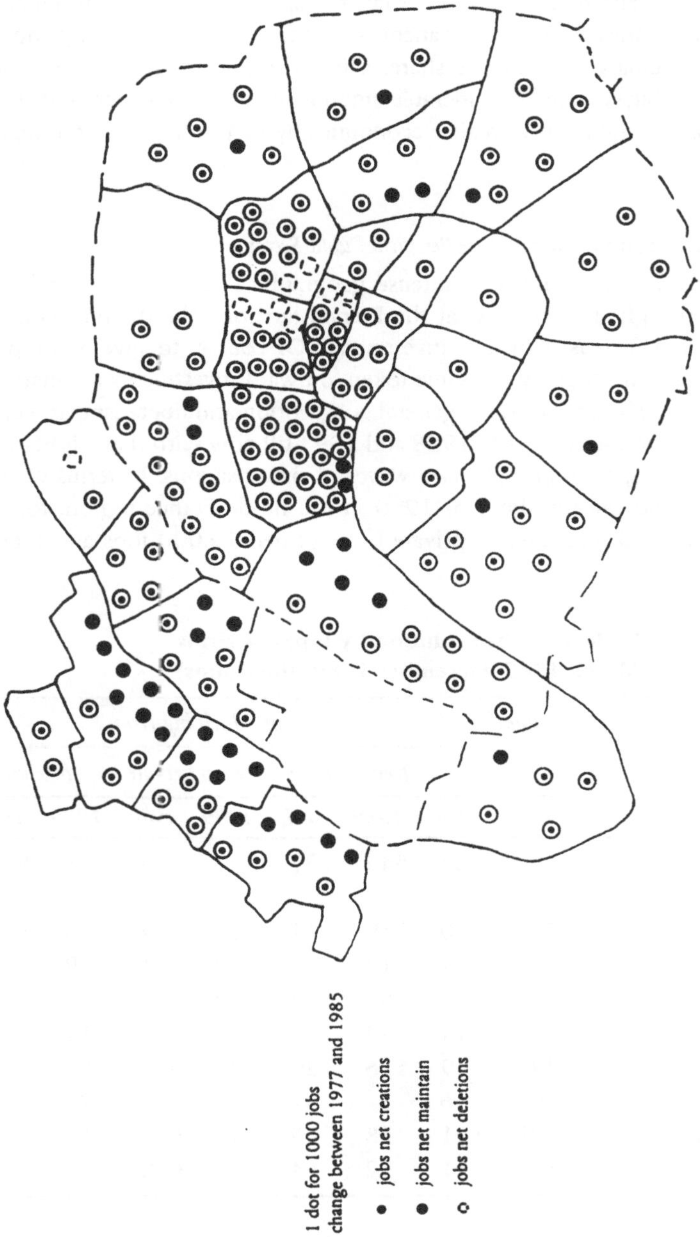

Source: elaborated from the 'Enquête sur la Structure des Emplois' data (After Cohen, J. (1993), in Métropoles en déséquilibre?, Paris, Economica, p.215)

***Figure 4.4** Producer-services jobs in the Ile-de-France region (1977–1985) (continued)*

This evolution is related to the internationalisation of production and markets. The competition becomes more and more acute. To be competitive, the firms have to minimise their costs and maximise their quality and efficiency. They focus on their core business and delegate the other tasks to specialists. If they cannot reach this competitivity, they face the permanent risk of failure or purchase by another firm willing to enlarge its market share. On the other hand, to make these acquisitions, the buyers need considerable amounts of available funds, and this is another reason for the firms to realise economies by cutting non-central departments.

Inward Investments, Privatisations and the Persistent Public Sector

This internationalisation causes an intense geographical concentration of firms and, not surprisingly, if one looks at the French territory, the disappearance of French firms and increased inward investments. Of course, to give a complete picture, one must add that, by the same token, outward investments are made by French firms also. The shift towards general competition and liberalism, as well as the two political periods of 1986–1988 and 1993 till now with their right-wing liberal governments, have brought two waves of privatisations. In terms of jobs between 1985 and the threshold of 1990 in the Ile-de-France region, foreign firms gained 43,000 jobs, French private firms gained 19,000 jobs and French

Table 4.4 Jobs evolution by types of firms, March 1985–December 1989 (thousands)

	*total**		*of which big firms*					
			french public		*french private*		*foreign*	
	1989	*evol.*	*1989*	*evol.*	*1989*	*evol.*	*1989*	*evol.*
Ile-de-France	2 250	-87	545	-27	527	+19	188	+43
of which:								
75-Paris	788	-110	219	-11	164	-4	46	+10
77-Seine-et-M.	126	+7	17	+1	29	-1	9	+1
78-Yvelines	205	+15	50	-4	59	+10	16	+4
91-Essonne	155	+14	32	–	37	+6	14	+3
92-Hauts-de-S.	444	-39	126	-20	103	-12	53	+13
93-S.St-Denis	200	-6	31	-2	53	–	23	+5
94-V-de-Marne	198	+21	38	+5	50	+10	14	+3
95-Val-d'Oise	134	+11	32	+4	32	+10	11	+4

Source: ESE

Premises over 19 salaried employees, private, semi-public and industrial public sector.

* moreover, the present table originates in data computing on the only premises where activity is indicated, so 5% of the total jobs are lacking.

public firms lost 27,000 jobs (table 4.4). During this time, the SMEs lost 120,000 jobs (Cohen 1994b).

In contrast to the general distribution, it is not Paris but the Hauts-de-Seine *département* that hosts the maximum number of jobs of the foreign firms. In proportion of its total, the Essonne *département* has a significant foreign presence, where IBM, at Corbeil, is a major employer.

One can see that, despite the privatisations, the public sector is still dominant in the Greater Paris labour market. Together with some banks and social financial institutions (pension fund, social security), power supply, public transport, air companies, railways and urban transport, still have a very strong presence and give the Capital region its biggest networks, that allow the huge metropolitan function.

What About the R&D jobs of Greater Paris in the 1990s?

I have shown that secondary-industry suffered an important decay. But the jobs lost were mainly in manufacturing (see table 4.2, p.55). The secondary-industry keeps numbers of jobs in the other functions, namely management, marketing and R&D. For this last, table 4.2 shows clearly that the share is higher in the west and south of Greater Paris than in the northern and eastern parts of the metropolitan departments (92, Hauts-de-Seine, west and south-west of Paris: 16% of the total secondary-industry jobs are dedicated to R&D; 78, Yvelines, western and south-western outer metropolitan area: 15%; 91, Essonne, southern outer metropolitan area: 12%; the north-western Val d'Oise, 95, reaches a 10% share, while Paris is lagging behind with only 3%, and Seine-et-Marne, Val-de-Marne or Seine-Saint-Denis stand in between with 7 or 8%). The Hauts-de-Seine *département*, that was still the most important in the region for secondary-industry jobs, keeps the first rank, and, by the same span of time (March 1985–December 1989), suffered maximum job losses and achieved the highest share of executives, engineers and technicians. Scrutiny of these figures clearly indicates the prominent part played by secondary-industry, even with a considerably reduced total of jobs, especially R&D, in modern productive systems.

An Assessment of the Present and Close Future of the Francilian Technopole

The Plateau de Saclay Among the Francilian Technopolitan Spaces: not only a 'Top-Down', but also a 'Bottom-Up' Development

The share of R&D among the industrial jobs of Essonne *département* (12%) is important, but it was still higher before: 13% in 1985. This is an unusually high share for an outer metropolitan area (Cohen 1994c). The reason is that this department had not previously had as many industrial jobs as the Hauts-de-Seine, Yvelines or Seine-Saint-Denis: no mechanical engineering industries, no car industry, not as much aeronautics. So, it did not house as many manufacturing workers as the northern and western departements and attracted only high-tech industries around a focus that can be localised at the Saclay Plateau, where biological research laboratories (1945, Gif-sur-Yvette), the Nuclear Studies Center (1952, Saclay) and the new Scientific University (1958, Orsay) settled first.

The historical framework of the settlement and development of this complex is well-known (Decoster and Tabariés 1992; Cohen 1994d). In 1936, the former 1935 Chemistry Nobel Prize winner Irene Joliot-Curie (with her husband Frederic) had been given a governmental post as State under-secretary for Research. She intended to create a new scientific university on a large campus out of Paris. Because of the coming second world war, this could not be done then, but it was realised in 1958. Irene Joliot-Curie probably looked to the southern suburbs as there was a metropolitan railway line linking the Parisian headquarters of the University of Science, the Quartier Latin, to those suburbs, reaching the Chevreuse valley (ligne de Sceaux, now RER B line). The first Centre National de la Recherche Scientifique (CNRS) research laboratories to settle in Gif were sent there by Frederic Joliot-Curie. In 1948, he and the French atomists, who were the guests of the French Army in the Chatillon Fort at Fontenay-aux-Roses, succeeded in their attempt to make the first atomic pile diverge. After that, in 1952, they settled in a considerably larger industrial estate at Saclay (the Nuclear Studies Center, CEN, of the Atomic Power Board, CEA), where they used to carefully organise the spin-offs of their inventions and works to enlarge the whole industry.

The first wave of industrial research settlements came at the end of the 1960s. Then French industrial firms, especially in electronics, having constituted a few international-scale companies and having dispersed their manufacturing and R&D all over France (especially in the former rural regions) concentrated their R&D in a scientific milieu. This attracted several engineering schools followed by other laboratories and scientific equipment suppliers creating new industrial estates, such as the Courtaboeuf Zone d'activites at Les Ulis. When these big companies began to suffer job losses (as military funding declined and the international struggle for scientific and technological progress was developing), several local actors (scientists and some prominent politicians, together with some firms' leaders, science park developers and so on), had the idea to strengthen and develop the links between science, research and industry. In this they followed the science parks model, like the Massachusetts Route 128, helping the local milieu to create its own jobs and enterprises. These sometimes tiny enterprises were not exactly comparable to the CEA, but this new type of co-operation to help industrialise scientific progress followed the previous customs of the CEA Nuclear Studies Center: to collect and develop spin-offs in a number of industries (metallurgy, surface treatment, robotics, biochemistry, biology or even spatial industry) and organise transfers of these innovations to the whole industry. One could even suppose that it will possibly not be as easy to ensure transfers and co-operation in a more commercial environment than it is within a CEA that has not only commercial concerns but also a public service role. As a matter of fact, one notes that it is easier for the biggest and wealthiest to collaborate, for example, for the Polytechnic School laboratories – that are basically funded by the Arms General Delegation of the Ministry of Defense – with companies like Thomson, Rhône-Poulenc or ELF, than it is for the less wealthy schools or university laboratories and SMEs.

However, the scientific potential of the area is really impressive. One local personality noted, for example, the weighing by a US firm needing numbers of mathematicians, between the french Saclay Plateau and Russia: this choice clearly indicates that the Saclay Plateau is at present one of the prominent concentrations of scientists in the world.

The New Saclay Plateau District: Does it Help Giving Sustainability to the 'Bottom-Up' Development of the Second Wave?

Despite the Decentralisation Acts of 1982–1984, vesting the territorial collectivities with new powers – including economic ones – the planning and urbanism guiding outline of the Ile-de-France region has remained under the responsibility of the State – which considers the Capital region an exception. This is the reason why, in 1988, the State asked the fourteen communes of the Saclay Plateau to join an inter-communal syndicate in order to prepare their own project. Given the general interests project, as established by the State, this inter-communal syndicate presented its proposition in 1990. In 1991 the district inter-communal du Plateau de Saclay was created, allowing the fourteen communes to take planning responsibility for their area – in keeping with the previously evoked planning and urbanism guiding outlines. Under this procedure, local politicians began to work with the scientists and economists involved in local development based on R&D.

Of the three development areas planned, two are currently growing. Le Moulon is the earlier but maybe the second, Palaiseau-Plateau, benefitting from greater funding (provided by the MoD rather than the Ministry of Higher Education and Research, or Ministry of Environment for the Moulon's project), will have the first new school transfer provided by the Ecole National Superieur des Tachniques Appliquees (ENSTA). There will not be a lot of buildings but accommodation should be supplied for temporary researchers (some five or six thousand beds are forecast). A small initial project is already under way. There will be an urban transport line linking the new town of Saint-Quentin-en Yvelines to the Saclay Plateau and Massy, which is an important communications node with two motorways, a TGV station, the B- and C-lines of the Express Regional Network (RER), and Orly airport. The preliminary discussions for the implementation of this link have begun whilst waiting for the decision of the Parisian Transport Syndicate (STP).

A lot of architectural and environmental projects are planned, more or less firmly. For example, the future 'Schema de secteur du Moulon' or 'Schema directeur hydraulique' for the water treatment. It is still too early to predict the results of these beginnings, but, if they are carried out, they will represent significant innovations.

Conclusion

In the 1960s with the increase in R&D jobs, in the 1970s with the constitution of big international-scale companies, and in the 1980s with the actions to help create innovative SMEs the French industrial policy seemed to be more and more committed to R&D. High-technology was the driving force of its economy. More and more synergies existed between university, research and industry. The public industrial sector helped in providing the whole country with infrastructure (power, transport, telecommunications), and also provided a good research jobs network all over the country. Despite the recent privatisations, this favourable disposition seems to persist. But since 1990, the industrial problems have returned and the true modernisation of the economy and the firms seems to have suffered more and more of this dangerous counter-part. The development of producer-services seems to have been slowed down by the persistent de-industrialisation. Induced by the international competition, the new commercial concern, even in the public sector, allows the firms to stay sound. But the danger is that R&D might have more difficulty in raising funding in the future. Presently, however, the high scientific level of the Saclay Plateau allows the local actors to be reasonably confident for the future of the technopole, and for its role in Greater Paris. But the development is still in its infancy and, at present, there is still a lack of agreement and co-operation in many cases – let alone the difficulty in planning during a time of recession. Equally, evolutions like price increases have some undesirable effects – as for instance, the difficulty for a greater proportion of the employees to find local accommodation. Undoubtedly the intrusion of considerably more market-led concerns is altering the original design of the Saclay Plateau development. It seems reasonable to retain the aim of a balanced development and to persist in helping both the scientific activity and the planning efforts.

References

Beckouche, P. and Cohen, J. (1992) 'Les etablissements de fabrication-stockage dans la premiere couronne d'Ile-de-France 1981–1989.' In S. Devoize (ed) *Les fonctions de fabrication-stockage dans la premiére couronne d'Ile-de-France.* Paris: IAURIF.

Cohen, J. (1991) 'Internationalisation, desindustrialisation, polarisations. La region-capitale se metropolise.' *Annales de la Recherche Urbaine 50,* 38–46.

Cohen, J. (1993) 'La nouvelle division intra-métropolitaine du travail dans le Grand Paris: centralites et complementarites fonctionnelles.' In METT-Plan Urbain, DATAR, CGP, avec le concours de l'Agence d'Urbanisme de la Communaut Urbaine de Lyon, Métropoles en desequilibre?, Paris, Economica, 199–219.

Cohen, J. (1994a) 'La desindustrialisation de l'Ile-de-France et ses composantes.' *Regards sur l'Ile-de-France 23,* 18–20.

Cohen, J. (1994b) Emplois et mutations du systéme productif: la métropole parisienne et l'Ile-de-France, paper presented at the IFRESI-CNRS-PIR Villes colloque 'Cities, firms and societies at the eve of the XXIth century', Lille, March 16-18, forthcoming ('Mondialisation et polarisation: dynamiques spatiales des emplois et mutations du systéme productif dans la métropole parisienne.' In A. Martens and M. Vervaeke (eds) *Polarisation sociale des villes europeennes.* Paris, Anthropos).

Cohen, J. (1994c) 'Industrie, emploi et extension urbaine: haute technologie et reseaux.' *Regards sur l'Ile-de-France 25*, 14–18.

Cohen, J. (1994d) *Technopoles de fait et amenagement: Paris-Sud et Londres-Ouest.* Paper presented to the European Research Symposium on Technopoles, Rennes, April 5–7.

Damette, F., Scheibling, J., Clech, L., Gresillon, B., Macchi, D., and Cohen, J. (1992) *Le Bassin parisien, systéme productif et organisation urbaine.* Paris: La Documentation Française.

Decoster, E. and Tabarés, M. (1992) 'Innovation and Regional Planning, the Ile-de-France-Sud Technopole.' In J. Simmie, J. Cohen and D. Hart (eds) *Technopoles Planning in Britain Ireland and France: The Planned Regional Acceleration of Innovation.* University College London, Planning and Development Research Centre, Working Paper 6, 83–101.

CHAPTER 5

What Comprises a Regional Innovation System?

Theoretical Base and Indicators

Heidi Wiig and Michelle Wood

Introduction

This chapter examines the innovation activities of manufacturing firms at a regional level, focusing on the county of Møre and Romsdal in central Norway. A key basis for research into regional innovation systems, exemplified by the great amount of attention this has received, is an awareness of the implications of disparities which exist between regions in terms of economic and technological growth and development (CURDS 1987; Landabaso 1995). By determining what distinguishes growth regions from less-dynamic regions, it may be possible to address the problems of those regions with less-developed economic and technological bases, which are often geographically peripheral regions.

The importance of innovation is also reflected in a shift of focus in regional policy away from purely economic issues and toward science and technology concerns (Logue 1995). The basic idea here has been expressed by the European Commission as: 'Regional economic performance depends upon the progressive introduction over time of innovations in products and processes to enhance the competitiveness of the regional economic base in an increasingly competitive world' (CEC 1991). This perspective has also emerged as a key issue in Norway. As policy makers are increasingly concerned with reducing regional disparities, this focal shift towards the science and technology aspects of regional industrial activity is reflected in changes in regional policy objectives (White Paper 1992–1993).

But, how should science, technology and innovation perspectives be incorporated into the analysis of regional economic performance? One of the key insights of modern innovation theory is that innovation is systemic, in the sense that firm-level innovation processes are generated and sustained by inter-firm relations and by a wide variety of inter-institutional relationships. Innovation and the creation of technology involve systemic interactions between firms and their environments: central links include those with customers and suppliers, science and technology infrastructures, finance institutions and so on. Such ideas have been central to the 'national innovation systems' literature (Lundvall 1992; Nelson 1993), which can be extended to the regional case.

A major problem, however, is to build an adequate empirical basis for conceptual work focusing on regional innovation systems. Much existing work is marked by the overall lack of comparable and comprehensive empirical evidence (Alderman and Wood 1994; Higgins 1995), and the absence of a developed theory that might provide a framework for further work in this area (Landabaso 1995; Higgins 1995). Although these problems highlight major objectives for future work, as part of our research we aim to contribute to a further theoretical and empirical understanding of regional innovation systems and to establish and test a research methodology that may be used in future regional innovation studies. As yet, however, there has been limited empirical evidence concerning regional technological diversities within Norway and existing studies from other countries are often of little relevance for the Norwegian case – mainly because of the special geography (spatially extensive, with many fjords, mountains and rural areas) and industrial base (often 'traditional' sectors) that exists there. Such factors have made it difficult to find directly comparable and comprehensive empirical analysis from other European regions. Accordingly, our survey is based on a structure and approach which has already been widely used to generate harmonised innovation data at national level in Europe, the approach of the so-called Community Innovation Survey. Our intention is to use this approach to start mapping differences between regions within Norway and, by developing an understanding of these differences, to suggest more effective and diversified policy measures. This study of Møre and Romsdal provides an initial step in this process.

This chapter is structured as follows: first, a brief background of studies of innovation at a regional level is undertaken in order to provide both empirical and conceptual bases for our research on Møre and Romsdal. Following this, an overview of the Møre and Romsdal region is provided, suggesting why it was selected for this study. This is followed by a more analytical discussion of technology-related issues associated with the regional innovation system, including an investigation of the region as a base for firms' innovation activities and the actual innovation activities and capabilities of firms located there. Specific regional factors which affect innovation activities of firms are examined. Finally, a summary and conclusions are drawn from the analysis, suggesting possible policy responses. An outline of the research methodology used is given in the appendix.

Conceptual and Empirical Bases

There have been long-standing efforts to understand, in theoretical terms, the economic and technological dynamics of industrial systems operating in particular regions (see Amin and Robins 1990; Asheim 1992; Brusco 1990; Castells and Hall 1994; Porter 1990; Storper 1991). The role and importance of geography or locational factors in this often forms a prime focus, where spatial proximity is a key factor in determining the outcome of firms' activities. This may enable the exploitation of 'dynamic relative advantages' of a given territory (Héraud 1994), arising from synergetic relationships between actors in the innovation system and economies of scale in the provision of innovation services and support. This is emphasised by Storper (1991) who states: 'Innovation and modification of products and processes...rests on an extraordinary complex variety of institutions, social habits, ideologies and expectations, and even firm and market structures are to a certain extent outcomes of these underlying social structures' (p.36) where the social structures are seen to be bound to specific regions. Alternatively, geography is present in analyses in the sense that studies are often explicitly or implicitly 'place specific' and base their concepts on observations from particular regions or localities where the innovation system is highly visible, for example, Silicon Valley in California, the 'Third Italy', Baden Württemberg in Germany (Cooke and Morgan 1994) and other 'innovative milieux' (Aydalot and Keeble 1988).

As a result of this, there has arguably been an over-emphasis on core regions and high-tech industries in the literature and the sporadic nature of such studies often results in inconsistency in the use of conceptual tools across different studies. This also creates difficulties for the application of findings from such studies of core areas – particularly when attempting to analyse innovation systems or innovation dynamics in other, less technologically-advanced regions or in low-technology sectors. Thus, studies on regional innovation often cite the lessons which may be learned from successful, usually geographically core regions, without fully concentrating on the endogenous capabilities of less-developed regions. This has important implications for regional policy, as Koshatzky (1994) notes: 'the activation and more intensive utilisation of endogenous innovation resources for regional development constitutes an important challenge for a technology-oriented regional policy' (p.1–1).

In addressing such issues of innovation capability, another major approach rests on the application of concepts which place less emphasis on geography and use ideas from evolutionary economics, systems theory and innovation theory, giving rise to the idea of systems or network models for mapping innovation (Lundvall 1992; Todtling 1994). These recognise that technology does not exist alone but functions as an integrated part of a socio-economic system, for example as a national innovation system (Lundvall 1992; Nelson 1993). Thus the context within which firms conduct innovation may be highly important and may be modelled by analysing the interrelationships between social, economic and technological systems at various scales. The various components and linkages within and beyond such systems or networks form the basis for analysis, and include: other firms, such as customers and suppliers; education institutions and research laboratories as sources of skilled labour and knowledge; government agencies as sources of finance, regulatory constraints and support for innovation; financial agencies such as banks or venture capitalists; and providers of business services.

However, placing less emphasis on geography can create difficulties, since the role of factors arising from the particular locality or region within which the system operates is ignored or, at best, explained by the 'embeddedness' of firms within particular cultural environments. This criticism has been raised particularly by Krugman and it has been suggested that 'recently, however, there have been certain developments within economics which may mark the beginning of a closer relationship with economic geography in general and regional development theory more particularly' (Martin and Sunley 1995). There remain, therefore, key questions concerning the role and importance of geographic factors in the operation of the social, economic and technological systems within a specific region.

In turn, the marrying of theoretical ideas with empirical work has also been problematic (Higgins 1995), particularly with the need to apply new and more sophisticated empirical indicators which has emerged with the recognition of the complexity of innovation (OECD 1992). The main existing science and technology indicators, namely R&D data, patents data and bibliometrics, are often irrelevant to regions characterised by 'traditional' industrial structures, large numbers of small firms and an absence of science-based industries and formal scientific institutions. The indicator problems following from this have also been discussed in Norway and considerable effort has gone into developing a wider range of official and unofficial statistics on innovation (Smith 1992; Smith and Vidvei 1992; CEC 1994). In addition to this, it seems clear that the study of innovation systems in particular localities or regions should be based on an integration of suitable innovation and regional indicators, using methodologies to allow comparisons across different regions (Alderman and Wood 1994; Nam, Nerb and Russ 1990).

The Statistical Approach

During the 1980s a number of independent research teams attempted to develop survey approaches to innovation which would widen the scope of statistical methods in innovation analysis (for an overview of such work, see Smith 1992). These surveys mainly attempted to collect data on new product development and on the firm-level activities which supported such development. In the early 1990s these approaches were synthesised by the Organisation for Economic Cooperation and Development (OECD) into a statistical manual which recommended a future 'standard practice' for the collection of such data. This approach was taken up by the European Commission, in a collaborative action involving DG-XIII (European Innovation Monitoring Initiative) and Eurostat, who implemented a 'Community

Innovation Survey' in all Member States in 1993/4; this survey collected harmonised data on approximately 40,000 firms. Simultaneously, the European Commission sponsored a study exploring the possibilities of extending this approach to a regional level (Alderman and Wood 1994). This project, known as ERIS (European Regional Innovation Surveys), was important background for the study reported here.

The Community Innovation Survey collected three broad types of data:

1. Economic data on new product introduction and sales, R&D and non R&D inputs to innovation, sales and employment.
2. Binary data on, for example, patterns of technological collaboration.
3. Ordinal data, asking firms to rank the importance of various information sources, obstacles to innovation, support measures, and so on.

In this study we use identical definitions and questions on innovation inputs and outputs to those of the Community Innovation Survey. However, we also adapted the questionnaire to reflect a range of locational issues – such as location of main suppliers and customers, roles of specific regional agencies, importance of specific regional infrastructural institutions and so on. The questionnaire was applied in two stages in mid-1994 to the gross population of manufacturing firms in Møre and Romsdal; it is, in effect, a census rather than a sample survey. In the first stage a postal survey was sent to all firms. In the second stage, all non-respondents were contacted by telephone and asked to complete a closely similar 'core' postal questionnaire. Only 110 firms declined to respond. However, a large number of firms (approximately 570) were either not relevant (that is, they had been misclassified as being involved in manufacturing production) or were impossible to contact. Approximately 300 firms failed to respond to letters and phone calls and there must be a strong supposition that they were out of business. We received a total of 399 responses, which represents a response rate of 78.4% of the firms who we succeeded in contacting, and 48% of the population including the 399 non-contactable firms. A subsequent non-response analysis was carried out with the 110 non-respondents, which suggested that there were no significant differences between respondents and non-respondents.

The Møre and Romsdal Region

The focus of this study is Møre and Romsdal, which was selected because it is a recognised region for innovation activities in traditional industries (Wicken 1994) and has a higher share of total industry employment when compared with the Norwegian average (Table 5.1), and had one of the countries highest numbers of patents in both 1982 and 1992 (Haug and Skorge 1994). As such, it may be termed a 'core' region in Norway. However, there are characteristics which distinguish it from other regions in Norway and core regions in other countries. In terms of gross value added, it is only the ninth largest in Norway (Figure 5.1).

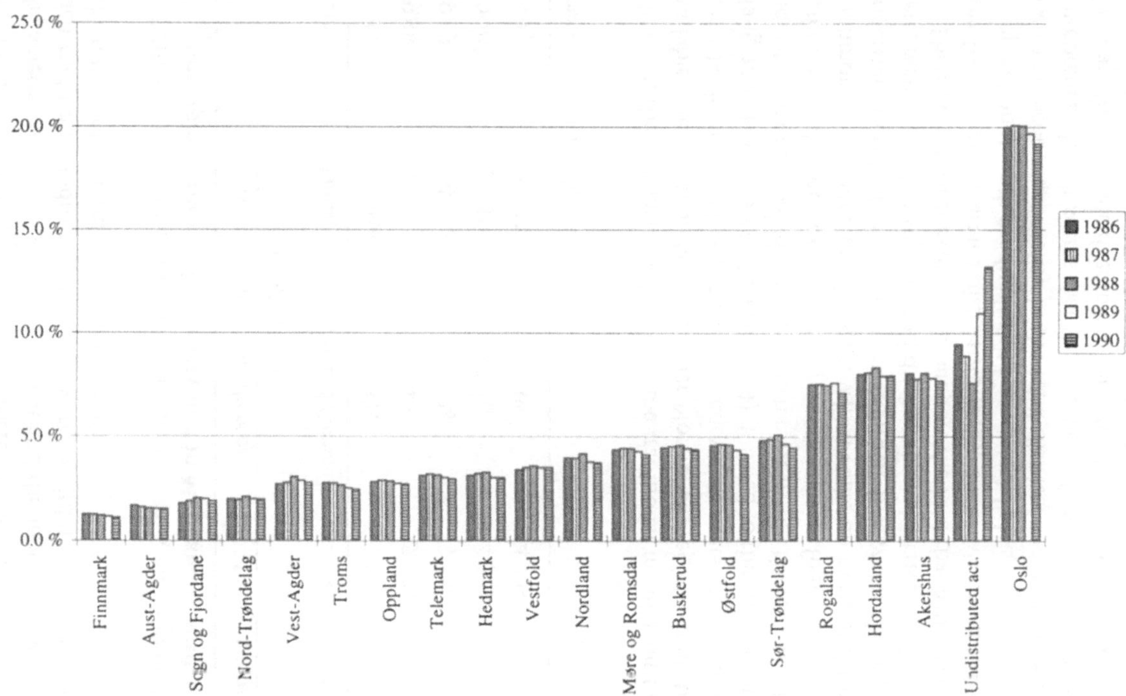

The methodological approach has been to allocate national accounts figures of gross value added (GVA) to regions by using distributional keys corresponding to each industry in the national accounts system.

Source: Statistics Norway

Figure 5.1 Regional gross value added at market prices. Million Norwegian Kroner (Total before deduction of imputed output of bank services, by county), 1986–1990

Also, there are the particular structural differences of the region, where the main industrial base is not founded on high-technologies such as electronics, computers and so on, but is largely comprised of three main 'traditional' industries – the manufacture of furniture, fabricated metal products (including shipbuilding) and fish products (see Figure 5.2). It is, therefore, useful to analyse the particular characteristics of the innovation system and the innovation activities of firms in such traditional industries, to see how these differ from other industrial sectors which have often received more attention.

Table 5.1 Industry employment as share of total employment and average unemployment rate in Norway and in Møre and Romsdal, 1987–91

	Industry employment as share of total employment		*Average unemployment rate*	
Year	*Norway*	*Møre and Romsdal*	*Norway*	*Møre and Romsdal*
1987	26.4%	30.4%	1.8%	1.8%
1988	25.9%	29.4%	2.3%	3.2%
1989	24.6%	28.5%	3.8%	4.7%
1990	24.1%	27.7%	4.3%	4.5%
1991	23.2%	27.3%	4.7%	4.6%

Source: Statistics Norway

Although the raw material bases for the traditional manufacturing industries are available often in abundance in other Norwegian regions, industry in Møre and Romsdal is perceived as particularly innovative in its use of these materials (see Haug and Skorge 1994; Wicken 1994). Related to this, the history of the region indicates that there are diversities which exist within the region, where innovation activities and industries differ across the three fogderi[1] or sub-regions of Sunnmøre, Nordmøre and Romsdal which make up Møre and Romsdal. Further understanding of differences in innovation activities within the region, as discussed here, may be used to support and direct policy objectives in this area.

There is historical evidence to support the idea that Møre and Romsdal as an area has entrepreneurial skills. Historically, collective entrepreneurship through co-operation in both productive and commercial phases of economic activity gave rise to an economic vitality in the rural districts and a positive attitude towards entrepreneurship. It seems that the tradition of collective entrepreneurship paved the way for individual entrepreneurs in these regions. The local community supported new enterprises by way of family, community or municipal support in terms of technical, financial and commercial support to initiatives taken by

1 An archaic jurisdiction akin to a bailiwick.

individuals (Wicken 1994). We suggest that these historical dimensions of the region may still be visible in the contemporary data.

Industrial Base of the Region

In 1994, 1128 companies were registered in Møre and Romsdal. Of these, by stripping out non-relevant firms or those no longer in business, 824 firms (representing 100% of manufacturing industry) formed the sample base for our survey. Using our primary data, together with information from the More and Romsdal Industry Catalogue 1991–92, the industrial structure is shown to be dominated by small, even micro-companies, since only about 100 companies in Møre and Romsdal employ 50 or more, whilst nearly 700 companies have less than 10 employees. However, when comparing this with the national pattern, it is noted that the average company size in Møre and Romsdal is in fact somewhat larger than in the rest of Norway. The share of companies in manufacturing, where the average company size is larger than in other economic sectors, also exceeds the national average. Thus, of all the 19 counties of Norway, Møre and Romsdal has the largest share of employees involved in manufacturing (see Table 5.1, p.72) and employment is to a large extent concentrated in a few industries, some of which are key to the national economy. These are:[2]

- manufacturing of fabricated metal products, machinery and equipment (ISIC 38), of which ship and boat-building (ISIC 3841) is the dominant
- manufacture of wood and wood products, including furniture (ISIC 33), of which manufacture of furniture and fixtures (ISIC 332) accounts for around 80%
- food manufacturing (ISIC 311–312), of which canning, preserving and processing of fish (ISIC 3114) accounts for 40–50%.

More specifically, at present roughly 40% of those employed in manufacture in Møre and Romsdal (Møre and Romsdal fylkeskommune 1993) work in manufacturing fabricated metal products, where ship and boat-building and manufacture of components and fixtures for ships and boats dominates. This industry has, meanwhile, experienced thorough restructuring and rationalisation, which has left waning demand for labour – despite satisfactory levels both of orders and profits. The furniture manufacturing industry in Møre and Romsdal now accounts for about half the sector nationally and was developed by individual entrepreneurs as a spin-off from traditional locally-organised woodworking activities. Today, the furniture industry is highly automated and its intensive use of technology has made it competitive both nationally and internationally. Despite this, only a relatively small share of production is exported. Finally, fish processing, together with the

2 We have used 'Standard Industrial Classification of all Economic Activities – ISIC' published in 1972 in grouping manufacturing industries. The classifications are given in parentheses.

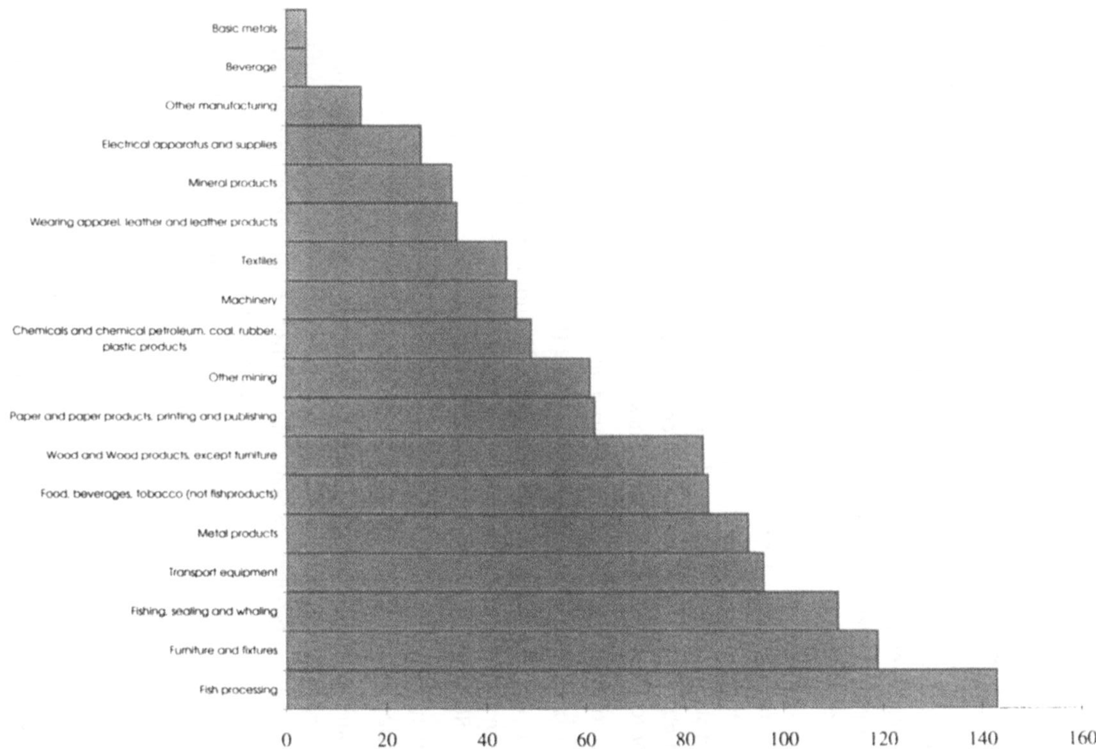

This is slightly less than the local number of firms in the region as we took out some non-manufacturing firms (retail and construction firms [6]) and the remainder were not allocated on 151c code. See appendix for fuller picture of sampling procedure

Source: M&R Advisory Council

Figure 5.2. Industry structure by sector in Møre and Romsdal, 1994. (n=1110)

fitting-out of the fishing fleet, may be viewed as an extension of the traditional base of activities that grew up around fishing. Modernisation of the region's industries may be described as a combination of local and international processes. While the fish industry is relatively conventional in the catching and processing of fish, for example, this can involve considerable technical upgrading which improves traditional technology; there has been important progress in research institutions which provides new possibilities for the industry. However, as yet, the fish industry has been unable to use these new inventions to their full potential (Hernes 1986). The fishing and the fish processing industries employ about 7% of the working population, spread amongst numerous, closely linked, small and medium-sized companies.

Economic Links and Innovation

How important is the region as an economic environment in input-output terms? It is evident from the survey results that there is a strongly focused economic base within Møre and Romsdal in terms of its importance as a market for firms' products, firms' links with key customers and as a source of supply for other firms. Customers within the region account for 61% of total sales and over half (53%) of the firms have their main customer in the region, trading mainly with industrial customers rather than supplying to final consumer markets. In addition, 34% of firms have their main supplier in Møre and Romsdal. These links are supported by the fact that the most important general regional factor affecting firms' activities in the region is presence of major customers or access to markets (Figure 5.3). This would imply a degree of local linkage formation in the form of 'clusters' (Porter 1990) or 'regional production networks' – particularly in the key sectors outlined above.

Innovation Inputs and Innovation-Related Expenditures

The survey evidence suggests that many firms are innovative in that they have expenditure on innovation activities, where 62% (n=249) of firms said they have some form of innovation cost (i.e. expenditure on innovation activities), although only 83 firms actually gave a distribution of total innovation costs (Figure 5.4). As shown in the figure, R&D expenditure (representing more basic or 'pure' research) accounts for only 12% of total innovation costs, whereas more applied work (the development side of R&D), such as trial production and product start-up (33% of total innovation costs), account for the majority of costs. In turn, developmental work and purchasing of products and licences are also key areas of innovation costs. This suggests that regardless of firm, R&D expenditure on the whole represents a relatively small element of the innovation process for firms in Møre and Romsdal, indicating that incremental innovations through learning by doing

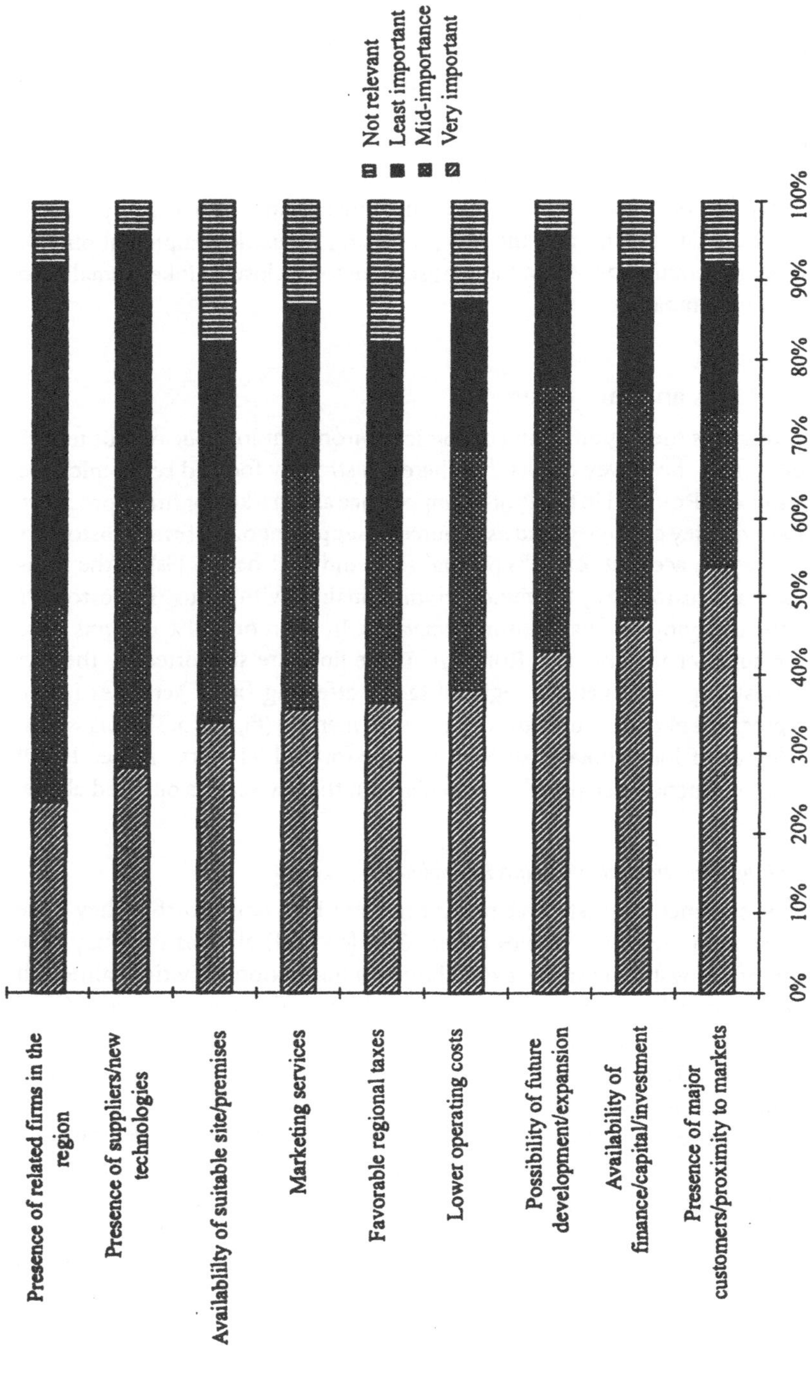

Source: STEP-Group survey of Møre and Romsdal

Figure 5.3 Importance of general regional factors to firms' activities (n=121), 1994

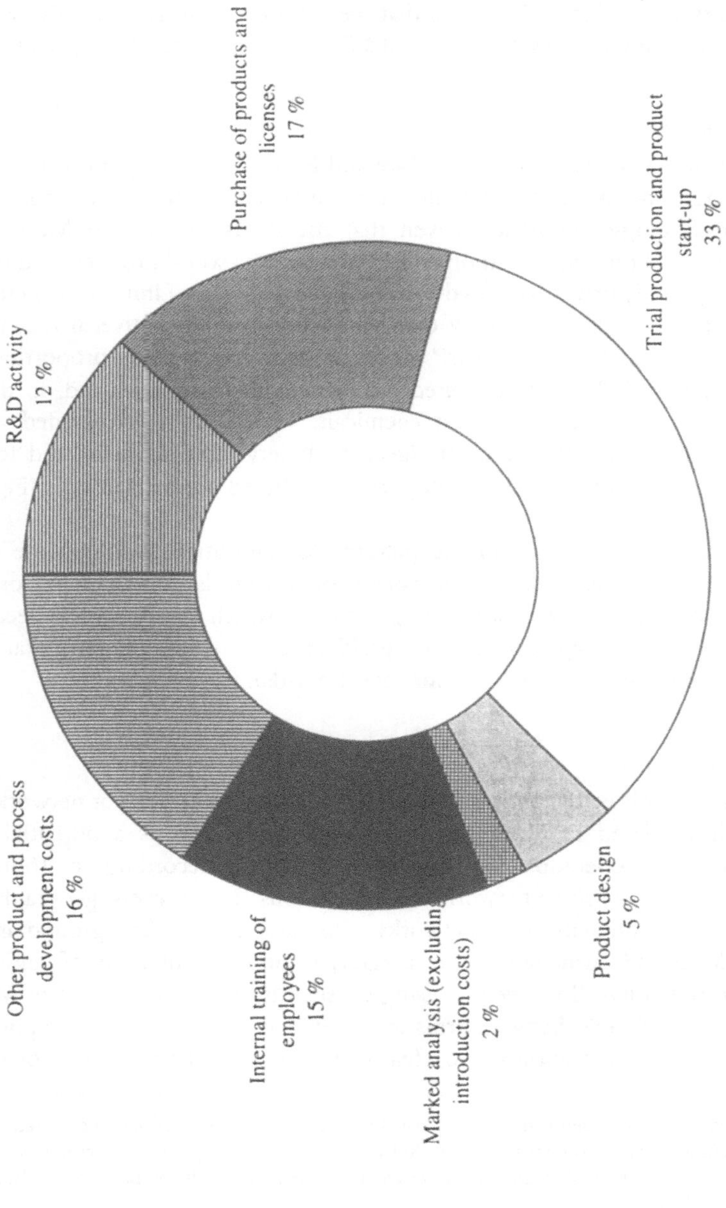

Source: STEP-Group survey of Møre and Romsdal

Figure 5.4 Distribution of total innovation costs (n=83), 1994

and learning by using are important in manufacturing industry in Møre and Romsdal. This (probably) reflects many small entrepreneurial firms in the region.

Firms also provided data regarding research employment. Of the 61 firms responding to this, only 26 registered full-time research positions – indicating that, overall, these firms have very few employees specifically engaged in R&D. These results are not surprising given that the majority of firms are SMEs, which often have limited resources directly for R&D expenditure and employment.

Innovation Outputs

The innovation activities of firms in Møre and Romsdal have significantly inter-industry variations and it should be noted that innovations are not only confined to 'high-technology' industries. Given that the industrial base in Møre and Romsdal is based on more traditional industries such as wood products and food products (especially fish) illustrates the persistent importance of innovation in these sectors. There is evidence to show that many firms are innovative in that they introduce new and altered products.[3] Wood products have a larger proportion of altered products (67%) than unaltered products in their turnover and, in these terms, is far more innovative than chemicals, which often is regarded as a 'high-technology' industry. 'Metal products, machinery and equipment' and 'food, beverages and tobacco' also have a proportion of altered products in sales (Figure 5.5).

There are important geographic patterns to innovation within Møre and Romsdal. For example, in terms of proportion of new products in sales, one of the three main *fogderi*, Sunnmøre, has the largest share – which may be largely because lack of access to risk capital is less of a problem for firms in Sunnmøre than for the other two sub-regions of Nordmøre and Romsdal.

Obstacles to Innovation

On the basis of this evidence, it is suggested that strong trade links or networking between firms based in Møre and Romsdal may be seen as a potential for interaction or co-operation for innovation activities. According to Tödtling (1994), for example, since networks exist at various spatial levels, geographical proximity, good communication networks, a common cultural background and a well-developed infrastructure act as a catalyst for the utilisation of regional innovation potentials. But when focusing on such issues in the context of Møre and Romsdal, such links between firms are not evident. In fact, the most important obstacle to firms' innovation is their fear of imitation of their products or risks

3 The main indicator of innovation output was the proportion of the firms' sales generated by product innovations introduced in the market within the last three years. Product innovations or 'new' products are understood here to mean either significantly altered products or slightly altered products.

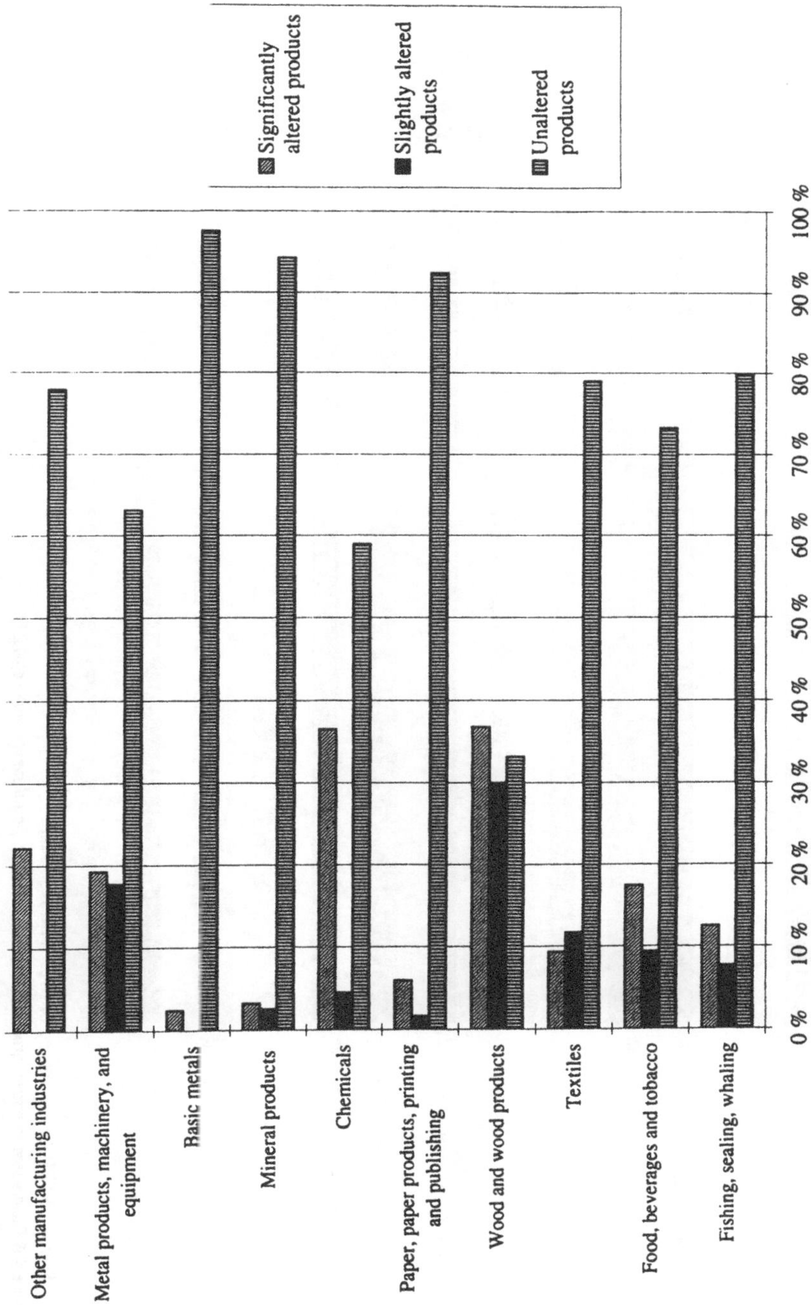

Source: STEP-Group survey of Møre and Romsdal

Figure 5.5 Share of turnover accounted for by unaltered, slightly altered and significantly altered products, by industry (n=252), 1994

Source: STEP-Group survey of Møre and Romsdal

Figure 5.6 Factors seen to restrict product/process innovation: postal survey only (n=121)

associated with being the first to innovate (Figure 5.6), this holds especially for small firms. This is further supported by the fact that firms see the presence of related firms of little importance to their activities (see Figure 5.3, p.76). Here again, especially for the small firms, the lack of co-operation possibilities is not seen as an obstacle to innovation (Figure 5.6). There are differences, in terms of size of firm, as to how they perceive lack of co-operation possibilities where larger firms see it as less of a problem than smaller firms. Such factors are more directly related to internal firm strategy rather than those based on regional factors, although, if there are a number of similar or competing firms in the region, the fear of imitation through, for example, loss of information due to spatial proximity between competing firms has an important regional dimension. This issue is highly relevant given the predominance of small firms in the sample which are often less able to support or finance risky innovation activities, indicated by the high response to high costs as a restrictive factor for innovation (Figure 5.6). This also reinforces existing evidence from other studies concerning the particular constraints to innovation faced by SMEs.

In addition, although previous research emphasises the importance of user-supplier interaction for innovation (see, for example, Lundvall 1988) this is not borne out by the evidence for Møre and Romsdal. Several firms are, in fact, dependent on one main customer (23% of firms rely on their main customer for more than 50% of sales), but the innovation potentials of such 'customer dependent' firms are lower (14% of innovation in sales) than those that are non-dependent (17%). Although the difference is not great, it does suggest that strong economic links with key customers does not necessarily have a positive effect on innovation. It shows 'the weakness of strong ties' (Grabher 1993), whereby such firms undertake a sub-contract role and are 'tied in' to supply customers with specific components or materials; as such they may have little requirement to innovate. However, co-operation for innovation between the firm and its main customer is not evident – suggesting a dependent supplier characterised by low technical skills producing only ordered components, which has a strong price competition because the customer has many related suppliers (Asheim and Isaksen 1995).

Overall, firms in Møre and Romsdal have strong trading links to the region; the goods that are sold are mostly low-technology but innovative, and most products are traded with industrial customers, rather than supplying final consumer markets. Firms are innovative and the fear of imitation suggests that there is high competition between firms in the region. There is historical evidence of entrepreneurial skills in the region and there are spatial differences in Møre and Romsdal when it comes to innovation, where the innovative regions have less problem with finding risk capital for their innovation activity. However, the results also indicate that too strong economic dependencies between firms does not necessarily promote innovation linkages between firms.

Role of External Links

Given the increasing awareness of globalisation (Howells and Wood 1993), expansion of export markets and emphasis on external technological collaboration (Chesnais 1988), the importance of external links to regional innovation has been emphasised. As such, firms' links both elsewhere in Norway and outside of the country were also examined in this study.

In terms of extra-regional trade links, 40% of the total sales of Møre and Romsdal industry are outside the region (Figure 5.7); the smallest firms (less than 10 employees) export only 14% of their sales, and for the largest firms, 58% of sales are exported (dominated by the EU market).

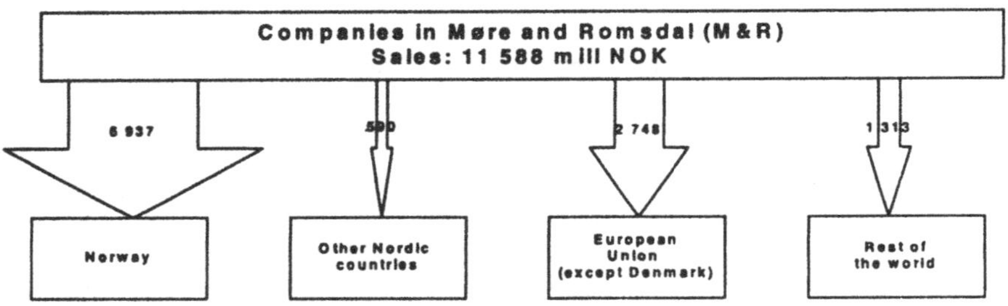

Source: STEP-Group survey of Møre and Romsdal

Figure 5.7 Sales from companies in Møre and Romsdal to different markets (n=350). (Figures in millions of Norwegian Krone)

In terms of markets elsewhere in Norway, the neighbouring regions to Møre and Romsdal – Trxndelag, Sogn og Fjordane and Hordaland – all have minor importance (accounting for 10% of all sales) and there is slightly more emphasis on markets in the Oslo region (12%) and elsewhere in Norway (17%).

As well as differing in terms of size of firm, with larger firms being more outward-looking, there is variation according to industrial sector. Thus the basic metals and metal products, chemicals, food, beverage and tobacco sectors are more export-oriented than other sectors (Figure 5.8), although in terms of national links, textiles and wood/wood products sectors are also relatively externally-oriented, contributing to domestic consumption needs within Norway.

Looking more specifically at innovative products in international trade, firms were asked to estimate what proportion of their exports in 1993 was accounted for by altered and unaltered products. Of the 40 firms that reported international trade, only 20 said that their turnover included altered products. Thus it appears that very few firms have innovative products amongst their exports. This may be because firms do not rely on export markets and are satisfied by supplying only regional and domestic needs. Conversely, there may be a lack of awareness amongst

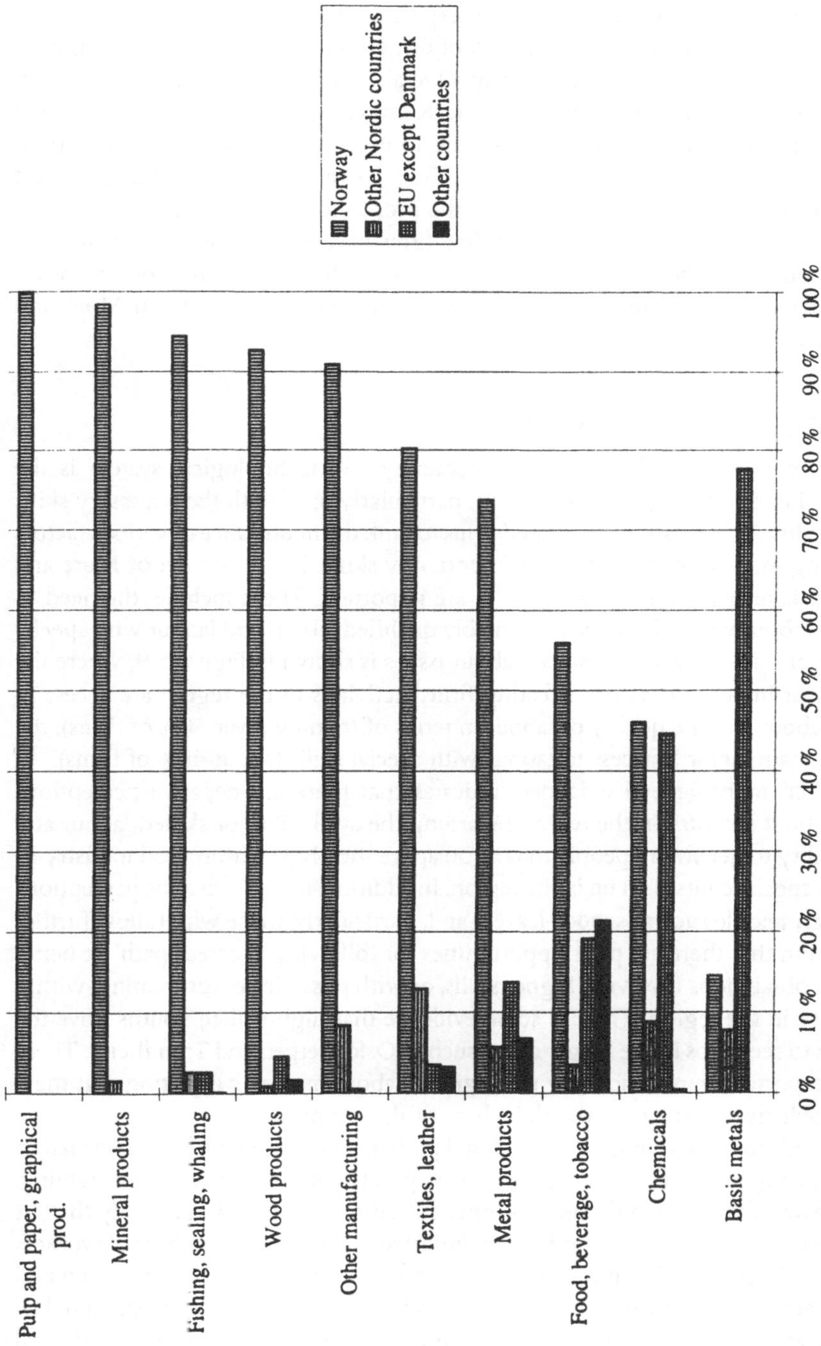

Source: STEP-Group survey of Møre and Romsdal

***Figure 5.8** Share of sales to different markets by industry (n=350)*

firms as to the potential openings for product innovation associated with gaining access to wider, and more competitive, export markets.

Investigation of external ownership of firms indicates that only two firms have parent companies outside of Norway. Overall, there are limited foreign direct investment links with firms in Møre and Romsdal – implying that international competitive pressures for indigenous firms to innovate, arising from proximity to foreign-owned firms, are limited. Thus, although overall firms in the region export 40% of total sales outside the region, the proportion of export differs between size and sector of firms. In addition, few firms have innovative products in their export market. Other external links, such as ownership relationships or via inward investment contacts, appear to be extremely limited within firms in Møre and Romsdal.

Availability of a Skilled Workforce

A key requirement of any regional economy and technological system is the availability of labour (IFO 1990), and, particularly, staff with the necessary skills and quality. The questionnaire asked firms to rank the importance of various factors affecting innovation activity, most importantly skills. In the context of Møre and Romsdal, several labour-related issues are important. These include: the need to obtain labour with relevant skills, suitably qualified labour and labour with special technical skills. This emphasis on labour issues is shown in Figure 5.9, where the two most important factors affecting firms' activities in the region are access to local labour and the quality of labour in terms of training (over 50% of firms); the fourth main factor is access to labour with special skills (about 49% of firms).

Other, more general evidence indicates that there are negative perceptions throughout industry in the region regarding the availability of skilled labour and the ability to get young people to take on apprenticeships within local industry or to take special courses set up in the region. In addition, there seem to be perceptions amongst people such as school-leavers and, particularly, those who follow further education that there are poor opportunities for following a 'career-path' or being able to obtain jobs involving higher skills, or with possibilities for training, within industry in the region. There is some evidence that high-skilled youths leave the region to seek jobs in the bigger cities, such as Oslo, Bergen and Trondheim. These have important implications for the regional labour market, suggesting that there are problems for renewing the skills base of the region.

In addition, the changing educational preferences of young people have led to a decreasing proportion of secondary level pupils going into vocational training. The share of young people taking higher education has risen dramatically the last couple of years and the labour force in industry is changing. The share of workers having a background in mechanical – and electro-engineering has decreased and the share of workers having administrative, economic, social science and law qualifications has risen. Regional industrial activity which is strongly oriented

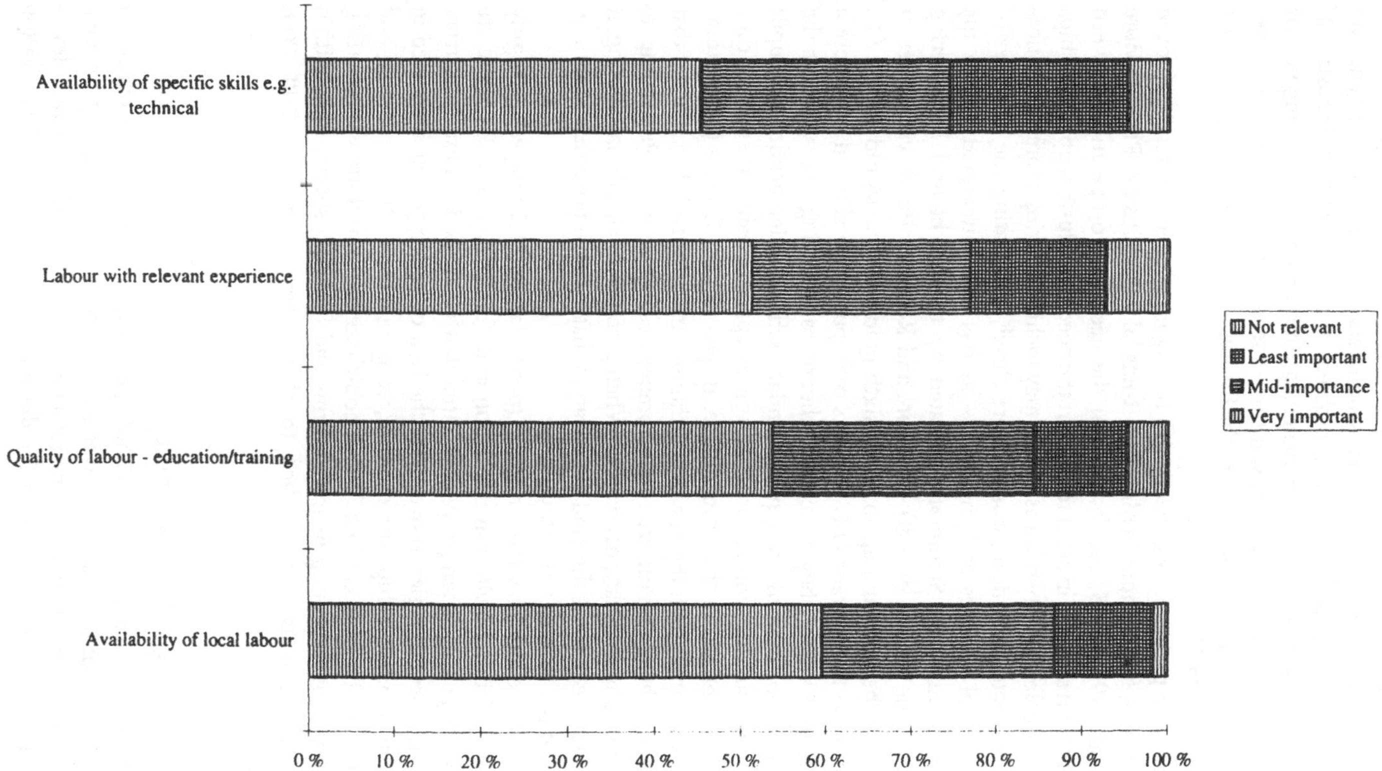

Source: STEP-Group survey of Møre and Romsdal

***Figure 5.9** Importance of regional labour factors for firms' activities: postal responses only (n=121)*

toward natural resources, and thus demands a high vocational intensity, has become less attractive for young people as increasing numbers are opting for more generally oriented education. This has been a problem for the furniture industry of Sunnmøre; an industry which has experienced great problems in recruiting youths for apprenticeships. "Today's youths want to educate themselves and 'be somebody' – if we want to attract the youth to the furniture industry it must get a higher status" (Aftenposten 1994).

There are also problems associated with having insufficient apprenticeships in certain industries, for example, in Møre and Romsdal. In the District Plan for Møre and Romsdal 1992–1995, the focus is on the need for co-operation between industry and high schools, in order to support the needs in industry apprenticeships that should be offered in these areas. These may include offering more apprenticeships with a greater technical basis or more closely integrating the activities of schools, technical colleges and industry – as has been attempted within the shipbuilding industry in Sunnmøre (initiated by the Mechanical Engineering Association in the Ulstein district) – in Møre and Romsdal. Pupils visit different shipbuilders in the third and again in the sixth grade (8/9 years old and 11/12 years old). In the ninth grade (14/15 years old), pupils are allowed to choose a course at school where they have both theoretical teaching and work at the shipyard and supplier firms. In upper secondary school (videregende) the pupils who choose shipbuilding will automatically get an apprenticeship, after which they are eligible to continue further into college. In general, the County (Fylkeskommune) wants to strengthen the vocational training in Møre and Romsdal, so that the industries in the region can raise their competence level and be able to compete on the national and international market. When it comes to vocational training, it is also necessary to provide the kind of education that makes it possible to continue higher education, such as university studies.

As the evidence indicates, labour factors (particularly skill shortages) are highly problematic to firms' activities in the region and, for many industries, there are problems with getting younger people interested to work in these industries. However, there are strategies to overcome the lack of young people interested in following an apprenticeship: for example, forging links between schools and industry at an early age (i.e. visits from schoolchildren to local industry), setting up apprenticeships for young people in firms and fostering special links within certain sectors such as ship building to allow young people to gain work experience.

Regional Technological Infrastructure

The role and importance of the technological infrastructure in the region in providing support for firms' activities was also examined. Geographical proximity, good communication networks, a common cultural background and a well developed infrastructure act as a catalyst for the utilisation and regional innovation potentials (Koschatzky 1994). Many firms in the survey indicated that more general infrastructural provisions related to the quality of telecommunications and proximity to key transport links are important. Of the firms, 38% perceived quality of telecommunication as the most important regional infrastructural factor to firms activities and 61% looked upon frequent and reliable transportation services as most important (Figure 5.10). The importance of these factors to the technological infrastructure is evident from many other studies in this field and, in the case of Møre and Romsdal, this is largely due to the special geography of the region, where towns are on different sides of fjords or mountains, so transportation links are of the utmost importance for firms.

However, although basic infrastructural factors are important to firms in the region, other technology-related factors such as proximity to higher education, technical colleges and research institutions are not perceived as important to their activities; this is especially true for the small firms, where 70% look upon this as least important but only 45% of the largest firms have the same view. There are no particular differences between the industries in how they perceive proximity to higher education, technical colleges and research institutions. Other evidence shows that 70 firms have been in contact with, for example, the research institution in the region (Møreforskning). We found only 11 of these firms among our respondents, of which 7 firms are seen to be innovative according to the criteria discussed above. In addition, the size distribution differs from our sample in that there is a majority of large firms that have been in contact with Møreforskning, 5 of these 11 firms have more than 100 employees but only 2 firms had less than 20. Thus, even though they have been in contact with Møreforskning, most of these firms responded to our survey as seeing proximity to research institutions as least or mid-important for their activities. This emphasises the results found earlier in the paper that showed that firms' R&D expenditures are only 12% of total innovation costs. This confirms that 'incremental innovations' through internal activities, or in co-operation with other firms, are important for the firms in the region.

More specifically in technological terms, it is evident from elsewhere that links between industry and the external technological infrastructure, such as universities and HEIs, are beneficial for innovation activities (Charles and Howells 1992). For firms in Møre and Romsdal, in terms of public support for innovation, the most important 'formal' source of support in the region is what is known as 'Møre and Romsdal firms' counselling' and the 24 regional offices of the State Industrial and Regional Fund (SND) (Figure 5.11). These institutions are regionally based and are acquainted with the region, organisations and institutions located and operating there and any barriers to innovation which may exist. As such, these institutions seem to have a positive effect on the establishment of links between firms and technological infrastructure. Other organisations, such as higher education institutes (HEIs) and technical colleges and schools, which could potentially offer a source of technological support and expertise for firms located in Møre and

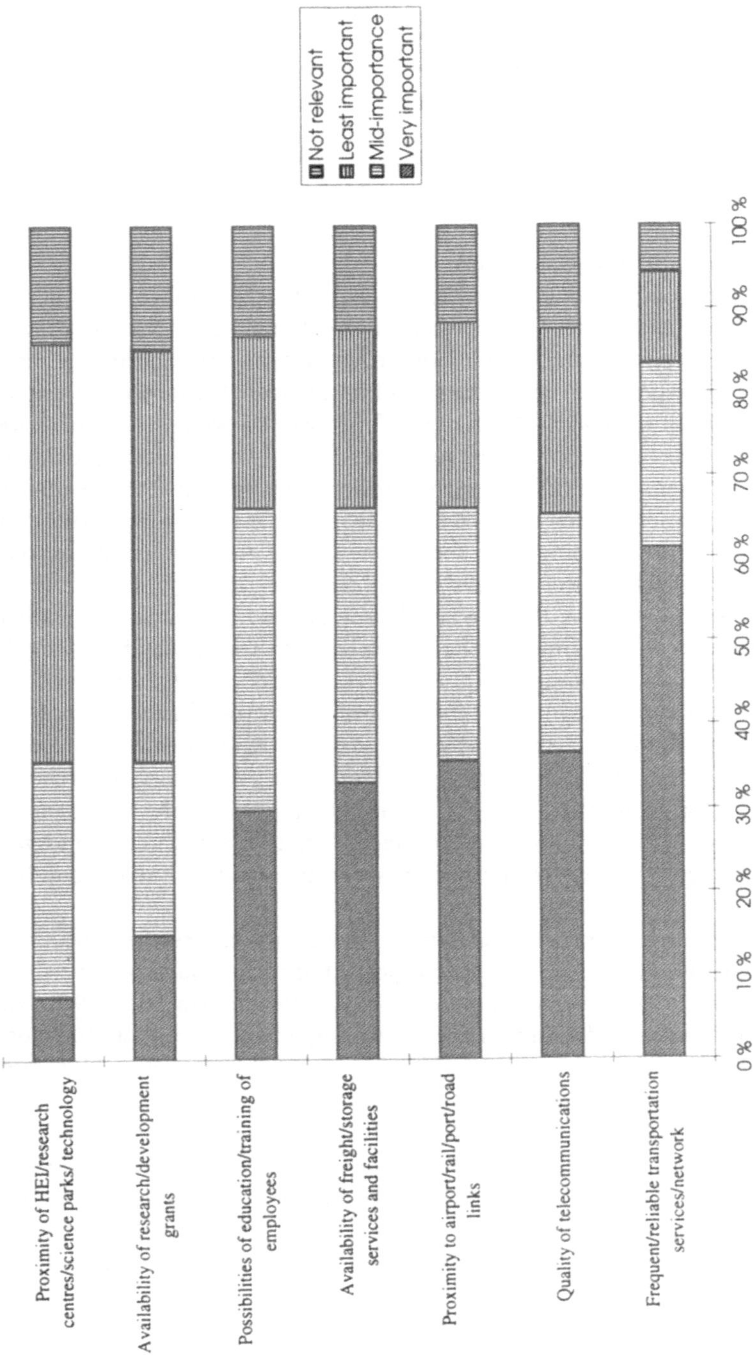

Source: STEP-Group survey of Møre and Romsdal

***Figure 5.10** Importance of regional infrastructural factors to firms' activities: postal responses only (n=121), 1994*

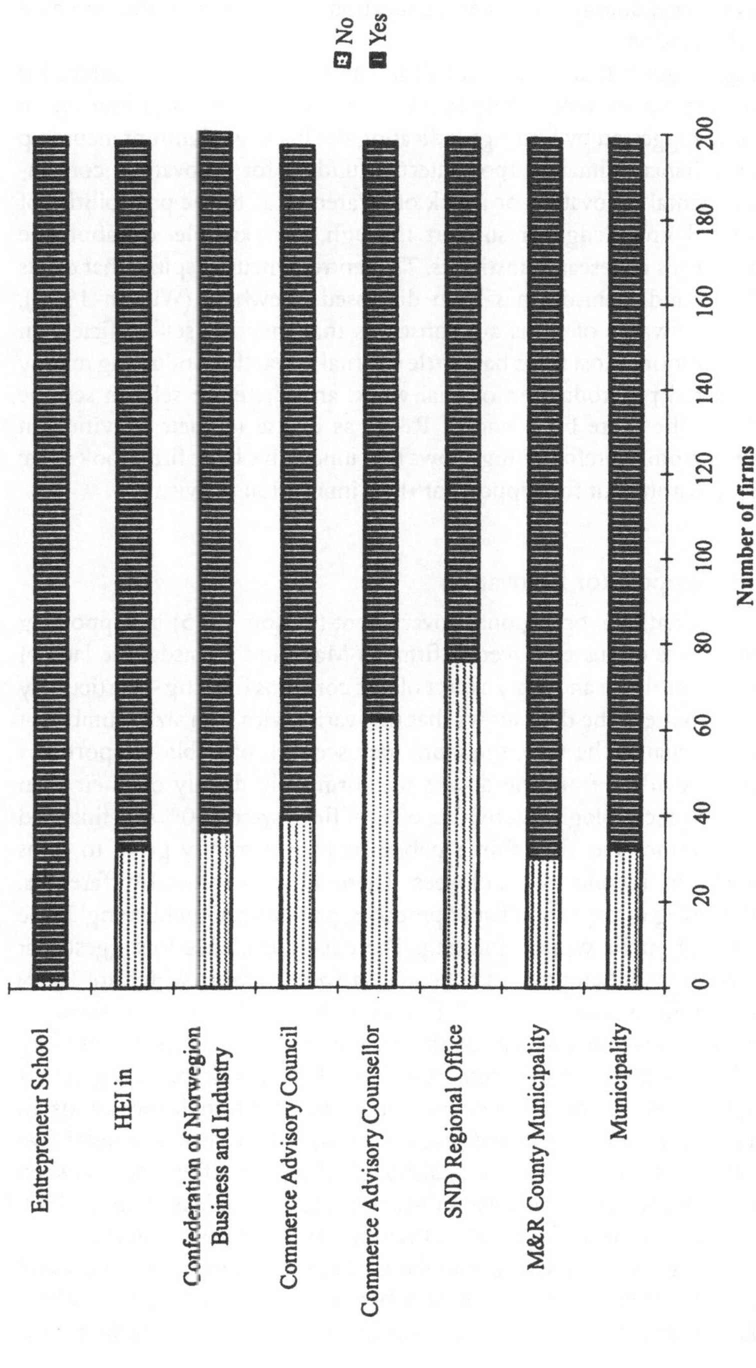

Source: STEP-Group survey of Møre and Romsdal

Figure 5.11 Sources of information, expertise or support for innovation activities: responses to telephone survey (n=121)

Romsdal, are not perceived as strongly important by firms. This again may be due to a lack of awareness amongst firms and education institutes as to the potential benefits of collaboration.

Our findings suggest that firms place little emphasis on external sources for innovation support but are more likely to rely upon internalisation of innovation processes. This is suggested by findings indicating the history of entrepreneurship amongst firms, reliance primarily upon internal funding for innovation, concentration on incremental innovation or a lack of awareness as to the possibilities of obtaining external knowledge or support through, for example, collaborative projects with colleges or research institutes. The 'entrepreneurial spirit' that exists in parts of Møre and Romsdal has been discussed elsewhere (Wicken 1994), implying that the owners of firms are conscious that they are self-sufficient in undertaking innovation. Most firms have little internal research, conducting mainly development-related preproduction or trial work, and therefore seldom see the work of HEIs (at the more basic end of R&D) as of use to their activities. In addition, the data from Møreforskning shows that innovative large firms look more to the external environment for support for their innovation activity.

Role of Public Support for Innovation

Finally, the role of national or regional government (Callon 1995) in supporting innovation in firms is discussed. Overall, firms in Møre and Romsdal see lack of public support as a problem and a key aspect of this concerns funding – particularly for innovation activities. The data shows that this varies with firm size (number of employees) – the smaller the firm, the more they see lack of public support as a problem. What is evident from the data is that firms rely mainly on their own sources of funds for technological activities, 63% of firms were 100% self-financed in terms of innovation. The data shows public support is mostly given to firms which have between 10 and 49 employees. There are also sectoral differences, where 'manufacturing of paper and paper products; printing and publishing' is the highest recipient of public support funding. There is no evidence to suggest that firms which receive public support are more innovative than those which are 100% self-financed, the firms that were self-financing had 25% of their turnover accounted for by innovations whilst for those firms in receipt of public funding the level was 20%. Although the difference is not substantial, it may suggest that self-financing firms are compelled to be more innovative with their own funds. In fact it appears that lack of finance and investment capital are seen as a restrictive factor on process or product innovation, although there are differences between the localities where the most innovative sub-regions, for example Sunnmøre, look upon lack of finances as less of a problem than the less innovative regions.

In addition, there is not a strong correlation between innovation inputs and innovation outputs. There are also time-lags between inputs and outputs where innovation is concerned – particularly with small firms who introduce new products only intermittently, we should not necessarily expect a statistical link between innovation costs and innovation outputs in one time-period. Despite this, lack of capital remains an important factor in firms' innovation decisions, where over 20% of firms see this as highly restrictive (figure 5.6, p.80), over 40% of the firms see insufficient government support as most restrictive. This may be partly due to internal lack of funds, as well as a shortage of funds from other regional agencies or institutions. There are also different types of innovations costs across different industries and these vary according to size of industries. These factors have been relevant when making public financial support available to firms.

It is recognised that there are other complex mechanisms that have positive effects on innovation besides financial expenditures. Other sources of indirect public support may include particular legislative arrangements, taxes or subsidies and other local economic development strategies. Although the main focus here is on financial support (particularly funding for innovation activities, given that it is a key issue for SMEs), firms also emphasised that there is insufficient government support in general and, more specifically, a lack of information regarding research/technology programmes (Figure 5.6, p.80). In general, though, the main finding is that firms lack finance capital, although those that are given public support in the innovation process are not necessarily more innovative than other firms.

Summary and Conclusions

The main aim of this chapter was to discuss the possibility of there being a 'regional innovation system' operating within and beyond the Norwegian region of Møre and Romsdal. The findings are based on empirical evidence from a survey focusing on the activities and responses of manufacturing firms, including: their economic and innovation activities, and links both within and beyond the region; the availability of labour, and education and training requirements; the technological infrastructure, including links with innovation support organisations; the role of public bodies in providing support for innovation.

The results show that Møre and Romsdal is an important base for the economic activities of firms located there, where the majority of trade occurs between firms within the region. Additionally, according to our evidence, many firms actually undertake innovation in products and processes. There is a strong regional economic environment and a specific type of innovation system; the question is, whether it is a dynamic system in terms of user-producer interactions. The existence of strong trade linkages and the presence of a number of firms, particularly in the three main industrial sectors, implies some form of 'clustering'. This suggests that benefits may be achieved via collaboration between firms, together with other institutions, for innovation in products and processes, as well as in the provision of trained labour, collaboration for provision of services and common technological expertise.

However, there is little evidence to suggest interaction between firms for innovation; in fact the presence of related firms is seen as unimportant to firms' activities and firms do not see other institutions as valuable sources of information, expertise and support for their innovation activities. In the main, they look to particular regional agencies, such as State Industrial and Regional Fund (SND) or the regional office for industry, for support, primarily in the form of funding, and appear to rely on internal entrepreneurship for their innovation activities. In addition, firms in Møre and Romsdal appear to face particular problems related to a general lack of public support for firms' innovation activities, the availability and retention of skilled labour including training and education (marked by rising unemployment since 1987 – see Table 5.1, p.71) as well as the more general problems SMEs in traditional sectors face in relation to innovation activities. A further difficulty may lie with the locational peripherality of Møre and Romsdal within Norway and Europe in terms of distance from potentially new markets and suppliers of technologies and equipment, and potential exclusion from other external sources of technological expertise, support and funding (for example, EU-funded projects).

These issues could form a key focus for public sector support and policies. Since there are mainly small firms in the region, these have particular problems related to lack of specialist capacities, 'bounded vision' (for example, a lack of awareness of innovation possibilities due to low resource and knowledge bases and limited expertise) and often strong locational dependency. All of these characteristics can affect their approach toward innovation and may affect their attitude toward external sources of technological support. Thus firms may be constrained in their use of external sources of support for innovation due to a lack of awareness about innovations developed in related companies, industries and public institutions. Equally, they are likely to experience insufficient in-house resources to enable external linkages or may view these with suspicion. In addition, the key sectors in the region are regarded as traditional industries and although there is evidence to show that some firms are actually using new technologies (and possibly 'high' technologies), whether in products or processes, there is scope to further develop these sectors. For some regions, industrial structure makes them vulnerable to the effects of geographical distance to markets, key suppliers and services, and partners and collaborators. Firms in these regions experience higher barriers to gaining access to information, technology and knowledge that are relevant for their production. These barriers make it difficult for the firm to participate fully in technological development in the relevant markets because the functionality of various networks and channels are severed.

It is important for national and, more specifically, regional governments to be familiar with the particular needs of the firms in the areas for which they are responsible. It has been argued that 'the varying nature of problems facing small firms in different regions and the difficulties of addressing those needs with centralised policies' (Woodcock 1993) therefore requires a response from regional and local government. In the case of firms in Møre and Romsdal, regional policies should take into account evidence concerning firm's innovation activities, and inadequacies, or perceived unimportance of the existing technological infrastructure in the region. Public support must be directed to those aspects of the innovation process in which firms are actually involved, that is, product development and trial production rather than research, and therefore technical and business advice and support may be the most appropriate. This may require the creation of new institutions such as business support offices or regional technology agencies, or new mechanisms, such as partnerships between firms and other organisations and between government agencies within the region and out to the national and European or international levels. More importantly, public support should look to the promotion of collaboration between firms and existing, regional institutions such as colleges and schools. Such a role may lie with local institutions, such as 'Møre and Romsdal firms' counselling' and SND's[4] regional office, institutions that firms already recognise and use, albeit minimally. In addition, as different industries and different sizes of firms have different needs, this presents a potentially important role for sector-specific trade organisations, which work to link customers and suppliers vertically, in contrast to the more general 'horizontal' measures applicable for all industries.

More specifically, policies should address the current problems associated with attracting and retaining skilled labour and the training or education of young workers. At present, there is a tendency for many young people to move to the cities and to study at the universities of Oslo, Bergen and Trondheim, for example. Since these cities have a more diversified labour market, they become an increasingly attractive location for higher-educated personnel; for a large number of students, obtaining a degree increases the preferential barrier to moving back home. There is an important role for both central and regional government in this to ensure the forging of links between schools, colleges and firms through, for example, having people from industry on the board of technical colleges. Students must see there are options after vocational training, something that will make it more attractive to young people today. The education system must develop a flexible system that makes it possible to combine vocational and higher education, this might attract new students. Other locational factors that might attract (or retain) a skilled workforce in the region are 'soft' locality factors such as leisure facilities and housing (Koschatzky 1994). National government, together with public sector in the regions, may aid such a transformation by means of a wide range of initiatives related to the development of knowledge-bases and the acceleration of learning-by-doing.

---

4 SND – The Norwegian Industrial Regional Development Fund – has as its aim to estabish a profitable socio-economic environment for industrial development in Norway. SND helps product development and the establishment of new firms, and helps with the modernisation and readjustment of Norwegian industry. Employment in less favoured regions is of great concern to SND. SND gives loans, guarantees and economic subsidies.

A key question is raised as to whether such policy should be based on an indigenous growth strategy or, alternatively, if there should be increased emphasis on improving external trade and innovation links. If a strategy of improving or strengthening external links is required, there appears to be an important role for firms and regional agencies in attracting finance and investment capital. This implies that if firms want to strengthen their export links or move to new markets (particularly in light of the decision not to join the EU) or if home-based firms want to undertake exports, it seems that there should be methods of collaboration between different types of firms in order to increase awareness as to the possibilities offered by external export markets. Conversely, however, if a strategy of endogenous growth is to be followed, awareness of the possibilities of internal markets in the region and in Norway, and possibilities of innovation-led growth, may be required. Thus there are possibilities of focusing on the needs of the region in order to help firms contribute to endogenous growth and technological development.

The location of Møre and Romsdal in Norway is peripheral, particularly when considering the location and role of the major cities in Norway. These both attract economic and technological activities and are sites for the main institutions in the national technological infrastructure, such as universities, higher education institutions and research institutions. This fact constrains the more even spread or regionalisation of many science-based activities in regions such as Møre and Romsdal. However, the vulnerability of firms to the effects of geographical distance can be reduced by technological and physical infrastructure in the region that is up-to-date and functioning as part of an orchestrated national infrastructure. By creating and developing physical infrastructures, for instance telecommunications and transport systems which firms rate as highly important, the public sector can reduce some of the drawbacks of being located in a particular region and, by that means, augment the advantages of the location. Thus integration of regional infrastructures with national infrastructures has to be an important aspect of a policy having the objective of stimulating sustainable economic growth in the regions. Despite this, a national technology policy cannot usually take regional problem situations adequately into account, since neither its aims nor its instruments are adapted to regional particularities (Koschatzky 1994). In which case, further investigation of the characteristics and needs of other regions will give a more comprehensive understanding of the national system of innovation.

In terms of the innovation system in the region, the results discussed here provide only one, albeit important, perspective – that of manufacturing firms – and, as the discussion indicates, in terms of analysing innovation linkages, the 'innovation system' or network in Møre and Romsdal is limited in scope and extent. This supports findings from similar studies elsewhere (see Héraud 1995). It may be that such linkages are more highly evident amongst firms and institutions within a sub-region, such as Sunnmøre, giving rise to a 'local innovation system' or within particular industries, embodied in their particular knowledge bases or labour markets – although further evidence is needed in order to investigate this. In addition, the apparent weakness of the Møre and Romsdal innovation system may lie with conceptual difficulties and perceptions associated with models of 'regional innovation systems', resulting from studies of successful, core regions where innovation linkages are strong. In fact, the problem of constructing a definition of an 'innovation system' which is applicable to a whole range of different localities and regions has been raised elsewhere (Higgins 1995), with suggestions that individual firm strategies and networks actually work against the formation of a visibly integrated regional innovation system. This chapter attempts to provide an empirical basis for this, but recognises that a great deal of work remains in this area, in both conceptual and empirical terms.

APPENDIX

Methodology

In the regional innovation study of the Norwegian county Møre and Romsdal, data was collected via postal questionnaires sent to business enterprises in the region. This was followed up by a telephone survey (using a slightly modified questionaire) to all non-respondents to the initial postal questionnaire. The total number of responses was 399 representing a response rate of 49%.

Address lists covering all industrial firms in the county were supplied by the Møre and Romsdal Advisory Council. Constant cross-referencing with the Brønnøysund register, together with the Council's extensive first-hand contact with industry in the area, ensured that these lists were both comprehensive and up-to-date. In addition to firm names, these lists furnished the names of the managing director – which proved important when contacting the individual companies.

In all, the original list contained 1128 business enterprises. Some of the firms were not relevant to the study and were discarded, as in the case of individual companies simply constituting a division of an enterprise. The original list also consisted of firms that were not engaged in production and a number of the remaining industrial firms did not deem the study relevant to their activities owing, for example, to the fact that they were involved in closing down production. Further, some had changed their activities away from producton several years back, while still others had not been involved in production for at least three years. There were another set of firms that found the study irrelevant for them and businesses that had gone bankrupt.

Table 5.2 Individual elements of the data collection

Total number of firms on the list	1128					
Firms out-of-business		51				
Non-relevant firms			253			
Study's sample base				= 824		
Responses to the postal survey					140	
Responses to the telephone survey					259	
Declined to respond						110
Unable to contact						315

References

Aftenposten. (1994) 'Ikke fint nok å lage møbler.' 24 August 1994.

Alderman, N. and Wood, M. (1994) 'Surveys of regional innovation? A feasibility study for Europe.' *EIMS Publication no.9.* SPRINT, CEC, DGXIII.

Amin, A. and Robins, K. (1990) 'Industrial districts and regional development: Limits and possibilities' In F. Pyke, G. Becattini and W. Sengenberger (eds) *Industrial Districts and Inter-firm Co-operation in Italy.* Switzerland: International Labour Organisation.

Asheim, B.T. (1992) *Industrial Districts, Inter-firm Co-operation and Endogenous Technological Development: The Experience of Developing Countries.* UNCTAD Symposium on industrial districts and technology, Geneva November1992.

Asheim, B.T. and Isaksen, A. (1995), 'Spesialiserte produksjonsområder mellom globalisering og lokalisering.' In D. Olberg (ed) *Endringer i arbeidslivets organisering.* Oslo: FAFO.

Aydalot, P. and Keeble, D. (1988) *High Technology Industry and Innovative Environments. The European Experience.* London: Castells and Hall.

Brusco, S. (1990) 'The idea of the industrial district. Its genesis.' In F. Pyke, G. Becattini and W. Sengenberger (eds) (1990) *Industrial Districts and Inter-firm Co-operation in Italy.* Switzerland: International Labour Organisation.

Callon, M. (1995) *Recent Trends in French Institutions for Regional Innovation Policies: An Appraisal.* Presentation to NISTEP International Workshop on Regional Science and Technology Policy Research, RESTPOR '95, Japan, 13–16th February 1995.

Castells, M. and Hall, P. (1994) *Technopoles of the World: the Making of Twenty-first Century Industrial Complexes.* London: Routledge.

Charles, D. and Howells, J. (1992) *Technology Transfer in Europe: Public and Private Networks.* London: Belhaven.

Chesnais, F. (1988) 'Technical co-operation agreements between firms.' *STI Review 4,* 51–119.

Commission of the European Communities (1991) *Four Motors for Europe. An Analysis of Cross-regional Co-operation.* Fast Occasional Paper 241, CEC, DGXII, vol.17.

Commission of the European Communities (1994) *The Community Innovation Survey, Status and Perspectives.* CEC, DGXIII, Luxembourg.

Cooke, P. and Morgan, K. (1994) 'The regional innovation system in Baden-Wurttemberg' *International Journal of Technology Management,* vol.9, 3/4, 394–429.

CURDS (1987) *RTD in the Less-favoured Regions of the Community.* STRIDE Final Report, CEC, April 1987.

Grabher, G. (1993) 'The weakness of strong ties. The lock-in of regional development in Ruhr area.' In G. Grabher (ed) *The Embedded Firm. On the Socio-Economics of Industrial Networks*. London Routledge.

Haug, R. and Skorge, O. (1994) *Patenter i Norge. Økonomisk utvikling, bedriftsstørrelse og lokalisering som forklaring på variasjon i antall patentsøknader i Norge.* Siviløkonomoppgave Bodø Graduate School of Business May 1994.

Héraud, J-A. (1994) *Is There a Local System of Innovation in Alsace? An Analysis of the Firms Networks Based on an Empirical Study.* Paper presented at EUNETIC Conference, Evolutionary Economics of Technological Change: Assessment of results and new frontiers, European Parliament, Strasbourg, Oct. 6–8, 1994.

Hernes, G. (1986) *Fast i fisken? Fiskerinæringens markedsmuligheter, styringsproblemer og innovasjonsevne.* (The fish industries market opportunities, steering problems and innovation capabilities). FAFO, Oslo, June 1986.

Higgins, T. (1995) *The Spatial Allocation of S&T Assets and their Management – Measurement Indicators and Evaluation.* Presentation to NISTEP International Workshop on Regional Science and Technology Policy Research RESTPOR '95, Japan, 13–16th February 1995.

Howells, J. and Wood, M. (1993) *The Globalisation of Production and Technology.* London: Belhaven.

Koschatzky, K. (1994) *Utilization of Innovation Resources for Regional Development – Empirical Evidence and Political Conclusions.* Paper prepared for NISTEP Conference, International Workshop on Regional Science and Technology Policy Research, RESTPRT '95, Japan, 13–16 February 1995.

Landabaso, M. (1995) *The Promotion of Innovation in Regional Community Policy: Lessons and Proposals for a Regional Innovation Strategy.* Presentation to NISTEP International Workshop on Regional Science and Technology Policy Research RESTPOR '95, Japan, Feb.13–16th 1995.

Logue, H. (1995) *The Role of Research and Technological Development in the Regions.* Presentation to NISTEP International Workshop on Regional Science and Technology Policy Research RESTPOR '95, Japan, Feb.13–16th 1995.

Lundvall, B-Å. (1988) 'Innovation as an interactive process: from user-producer interaction to the national system of innovation.' In G. Dosi, C. Freeman, R. Nelson, G. Silverberg and L. Soete (eds) *Technical Change and Economic Theory.*

Lundvall, B-Å. (1992) (ed) *National Systems of Innovation.* London: Pinter.

Martin, R. and Sunley, P. (1995) *Paul Krugman's Geographical Economics and its Implications for Regional Development Theory: A Critical Assessment.* Paper presented at IBG Conference, Newcastle-upon-Tyne, January 1995.

Møre and Romsdal fylkeskommune (1993) *Årsmelding 1993.* (Annual report 1993). Nærings-og miljxavdeling.

Nam, Ch.W., Nerb G. and Russ, H. (1990) *An Empirical Assessment of Factors Shaping Regional Competitiveness in Problem Regions.* IFO Main Report, CEC, Luxembourg.

Nelson, R. (1993) (ed) *National Innovation Systems.* New York: OUP.

OECD (1992) *Oslo Manual. OECD Proposed Guidelines for Collecting and Interpreting Technological Innovation Data.* Paris: OECD.

Pike, A. and Charles, D. (1995) 'The impact of international collaboration on UK university-industry links.' *Industry and Higher Education.* October edition, pp.264–276.

Porter, M. (1990) *The Competitive Advantages of Nations.* London: Macmillan.

Smith, K. (1992) 'Technological innovation indicators: experience and prospects.' *Science and Public Policy 19*, 6, 383–392.

Smith, K. and Vidvei, T. (1992) 'Innovation activity and innovation outputs in Norwegian industry.' *STI Review, OECD, 11*, December 1992.

Storper, M. (1991) *Technology Districts and International Trade: The Limits to Globalization in an Age of Flexible Production.* Mimeo Grad School of Urban Planning and Lewis Centre for Regional Policy Studies, University of California LA, September 1991.

Todtling, F. (1994) 'The uneven landscape of innovation poles. Local embeddedness and global networks.' In Amin and Thrift (eds) *Globalization, Institutions and Regional Development in Europe.* Oxford: OUP.

White Paper (1992–1993) *By og land hand i hand* (City and district hand in hand) London: Pinter.

Wicken, O. (1994) *Entrepenørskap i Møre and Romsdal. Et historisk perspektiv.* (Entrepreneurship in More og Romsdal. A historical perspective) STEP-Report, 21/94.

Woodcock, C. (1993) 'A regional problem that needs to be addressed.' *The Guardian*, 5th April 1993.

PART III

Peripheral Regions

CHAPTER 6

Competitiveness and the Global Region

The Role of Networking

Robert Huggins

Introduction

This chapter examines the impact of increased networking awareness on models of regional development and competitiveness, with particular regard to innovation and technology policy. The study investigates the role of networks, specifically in the SME sector, which is often dependent on information and knowledge provision through external sources. The chapter focuses on the effects of networking at both a regional and global level, particularly the mobilisation of regions within the global economy.

The second part of the chapter concentrates on the role that an entity such as the 'technopole' can play in the regional economy and what might be its most appropriate make-up. The study examines a feasibility project which is currently taking place in Wales to design an innovation network, offering the opportunity for information and knowledge exchange at both the regional and global level. The 'South Wales Technopole' project is an EU-SPRINT funded study undertaken by the Centre for Advanced Studies in the Social Sciences at the University of Wales in collaboration with the Welsh Development Agency. The chapter relates the Welsh initiative to models of regional development in both Denmark and North-Rhine Westphalia, which have based their competitiveness and modernisation strategies on innovation support networks and the increased participation of their respective SME populations.

Competitiveness, Globalisation and Regions

As new modes and efficiencies in telecommunications and transportation make the world grow ever smaller (Reich 1991), the global economy, and its global markets, are increasingly viewed as the exclusive domain of power-hungry multi-national corporations. There is, however, a growing argument that globalisation is actually reinforcing the role that geographic clusters of production play in the competitive international arena (Porter 1990; de Vet 1993; Cooke 1994). Scott and Storper (1992) view the global economy as a mosaic of regional production systems, each with its own intra-regional markets and activities, and also the global web of inter-regional linkages. There appears to exist a paradox, highlighted by Castells and Hall (1994), in that although the technological revolution has provided the infrastructure for the increasing globalisation of economic structures, both cities and regions are becoming increasingly crucial agents of economic development. De Vet (1993) argues that regions in particular are becoming stronger economic entities through the fostering of growth via the mobilisation of their asset base, including their local firms. Such firms are increasingly interacting with the external environment and, according to Porter (1990), gaining competitive advantage through innovation, which he sees as improvements manifested in product and process changes, new approaches to marketing and distribution and new conceptions of scope.

As innovation is highly dependent on information and knowledge, these elements are becoming the critical success factors in new models of regional development (Nijkamp *et al.* 1994). A capacity to innovate appears prefiguratively to imply the necessity to access such 'invisible factors' through a networking capacity. This reinforces the argument that globalisation intensifies regionalisation, as information and knowledge is often of a global nature but needs to be delivered locally (Kogut *et al.* 1993; Nielsen 1994). Networking capacity at its most simplistic level can be seen as the disposition to collaborate to achieve mutually beneficial ends (Morgan 1994), but, as De Bresson and Amesse (1991) argue, networks have much in common with other powerful concepts in that they are necessarily all-encompassing and subjected to various usages and meanings.

Despite a lack of precise definition, network characteristics are seen to be of particular importance for innovation and technological change and the growth prospects of spatially defined regions (Bergman *et al.* 1991). Although such regional networks enable firms to tap into local expertise and knowledge, their true strength comes from their ability to interlink with other networks on a world-wide basis. Camagni (1991) views such links as imperative, pointing to the fact that the region, or 'local milieu', needs to be linked to international and global networks in order to stay innovative in the long term and avoid 'entropic death'. Also, as Christensen *et al.* (1990) argue, 'tight' networks in a regional frame can be seen as a preconditioning factor underlying the gain of competitive advantage in a global frame. These global-regional networks clearly have a different relevance for different actors (Tödtling 1994) and it is primarily small- and medium-sized enterprises (SMEs) who will have the most to gain. Consequently, it appears that regions who usually have less power than national governments, through the dynamics of regionally-based networks, provide an entry route for local innovative SMEs to global networks of information and knowledge. As the prevalence of mobile capital has increased the presence of large global companies, so networking capacity has facilitated the creation of global regions which are able to integrate geographically-restricted economies into the global web of industry and commerce. As Nijkamp *et al.* (1994) indicate, it is not surprising that many regions are increasingly reaping the fruits of a networked economy, in which regions play a central role in an international competitive system.

Whereas Porter's (1990) analysis of global competitive advantage sees the competitive unit as the industry or industry segment (Marceau 1994), the concept of global regions, by definition, puts the region at the centre of competition. There are two important factors which can be related to this concept: as accessibility loses its correlation to geographical distance, paradoxically, a global regionalisation of relations is occurring (Karlsson and Westin 1994) and as new technology, the key to innovation, creates more and more links between industries (Gomes-Casseres 1994), the competitive position of industries will become increasingly blurred – particularly in regions where there is already considerable cross-sectoral overlap. Interactions across industrial networks are particularly rich in information and knowledge, with synergies creating further knowledge and often resulting in dynamic technical accumulation and production improvements (De Bresson and Amesse 1991; Lakshmanan and Okumura 1995). Hilpert (1992) has argued that this is already occurring in Europe, and therefore European policies should support the linking of regions according to a cross-sectoral mode of innovation. Also, partly as a result of the Single Market, some argue that the main competitive entities in Europe are regions and multi-national firms (Amin 1992).

Network Models and Technopoles

The burgeoning literature on networks makes it impossible to synthesise all the relating issues, however, with regard to the global regions concept, it appears that there are three important and interrelated models:

- information networks
- knowledge networks
- innovation networks.

These network models become increasingly complex and, therefore, powerful, with information networks as the most simplistic and innovation networks being both the most interactive and difficult to implement. In between these are knowledge networks, which have increasingly come to the fore as a key instrument in regional development strategy. The presence of a knowledge network is increasingly regarded as a primary locational factor, particularly if it provides access to the broader international knowledge networks (Nijkamp *et al.* 1994). Successful global regions are those whose networks incorporate an adequate supply of quality knowledge resources, along with the ability and willingness of local firms to make use of external sources of knowledge with a clear focus on innovation (Nielsen 1994).

Kogut *et al.* (1993) argue that the structure of a network is more than a description of the flows of information and the differential availability of this information. It is an expression of the knowledge that influences the capability of individual actors, with more information leading to more relations. They distinguish between knowledge and information networks, arguing that:

> Knowledge consists not only of information...but also of the know-how regarding co-operation. This distinction between information and know-how operates at several levels of analysis, from the individual, to the group and organisation, and ultimately to the network itself. Information of the network consists of identifying who will co-operate and who has what capabilities. Know-how is the knowledge of how the capabilities of individual firms might be harnessed through co-operation. (p.77)

Therefore knowledge is the effective exchange of qualified information and presupposes communication or direct face-to-face contacts between individuals (Karlsson 1994). Similarly, Johansson (1991) distinguishes between knowledge and innovation networks, arguing that innovation networks proceed from information links between developer, user and other actors in a knowledge network whilst innovation networks revolve around the development of new technical solutions, new equipment and knowledge in a contact-intensive interaction. In other words, innovation networks originate from the creative combination of know-how through knowledge networks and specific skills (Maillat *et al.* 1994).

Much of the transfer through such networks inevitably involves technology. Charles and Howells (1992) distinguish three types of technology transfer which are comparable with the three networks models:

- information transfer – comprises data, documentation, software, standards, etc.
- knowledge transfer – requiring an understanding of the origins and potential impact of the technology or processes, skills know-how, and relevant policy issues and the ability to adapt and diffuse innovation
- hardware transfer – broadly interpreted as the transfer of devices, equipment, parts, materials and entire information systems.

As well as facilitating the emergence of networks, the knowledge revolution has been responsible for the creation of knowledge centres, which are at the heart of global regions. At the core of knowledge centres, or *technopolis* as they are termed, is usually an institution that creates new knowledge (Preer 1992). Technopolises have developed in regions that were primarily homes to centres of critical research and development and are the key nodes the of 'global knowledge network', generating new products and processes used throughout the world. As Amin and Thrift (1994) argue, these centres of knowledge are also:

> centres of representation, interaction, and innovation within global production filieres... In other words, these centres are the forcing houses for the

> construction of world-wide contact networks through which disclosures circulate and are modulated. (p.13)

Apart from technopolis, these centres are increasingly termed *technopoles*, after the French dirigiste science park tradition (Cooke and Morgan 1994). The technopole concept has generally been associated with attempts by central governments to implant innovative clusters in underperforming regions (Cooke *et al.* 1995). These attempts have more often than not failed as a result of the nodes of the technopole – universities, research institutes, training and technology transfer agencies, banks, firms, etc. – being isolated or non-networked, despite being in the same geographical proximity. Therefore, early technopoles were usually the consequence of:

> A hasty, hurried study by an opportunistic consultant...to provide the magic formula: a small dose of venture capital, a University (invariably termed a 'technology Institute'), fiscal and institutional incentives to attract high-technology firms, and a degree of support for small business. All this, wrapped within the covers of a glossy brochure, and illustrated by a sylvan landscape with a futuristic name, would create the right conditions to out-perform the neighbours, to become the locus of the new major global industrial centre. The world is now littered with the ruins of all too many such dreams that have failed, or have yielded meagre results at far too high a cost. (Castells and Hall 1994, p.8)

Castells and Hall (1994) argue that in many cases the technopoles have been constrained by boundaries artificially set by the promoters of the technopole idea. However, recent thinking on technopoles appears to have learnt from these failures, with increasing weight being given to the importance of networks and local consensus-building. One example where this approach is being taken is in Wales, where the South Wales Technopole project is focusing on what firms, as the likely users of innovation support, actually require and which is not readily available via market transactions.

The South Wales Technopole Project

The South Wales Technopole project was initially proposed by the Welsh Development Agency in 1992 through an application for the funding of a feasibility study to the European Union under SPRINT, its science park consultancy scheme, supplied by DG XIII. At that time, the objective of the South Wales Technopole was seen as being 'to establish a truly networked region as a new generation of "technopole" – a focus for innovation in the UK'. Funding was secured from SPRINT to undertake the feasibility study and also support a European Expert Panel to advise on, and evaluate, the study. It is interesting to note that although the Technopole concept is clearly different from the typical Science Park project that was previously supported by SPRINT, it has recently added a new funding category – Regional Innovation and Technology Transfer Strategies and Infrastructures (RITTS) – and the South Wales Technopole is seen as a model project of this type.

The rationale for the project was that, although the regional community in South Wales has significant scientific, technological and innovation resources, it lacked an effective means of communication to deliver the right information to the right SMEs. The solution to this problem was seen to be the creation of a regional innovation network linking firms (primarily SMEs), banks, technology transfer agencies, universities, government bodies, Training and Enterprise Councils, etc. in a web of information and knowledge exchange. The objective of this network would be to enhance the innovation and technology needs of SMEs in South Wales. This objective formed the context for the feasibility study which had the following aims:

(1) to assess the strengths of the existing innovation and technology infrastructure in South Wales.

(2) to determine the 'market' in South Wales for an innovation network, among, though not exclusively, SMEs.

(3) a study of existing European best practice innovation and technology networks.

An important element of the feasibility study was the convening of a number of public forums to disseminate the initial results of the study and to raise public awareness of the project. Such was the success of these public meetings that a group of small innovative firms, representatives of the Welsh Development Agency and two local TECs, as well as the Centre for Advanced Studies at the University of Wales in Cardiff, elected to form the 'South Wales Technopole Pilot Project' forum. The forum, which was held monthly from August 1994, and effectively became the Technopole steering group, sought to address some of the initial discussion issues which were raised by the European Expert Panel. These issues included the following:

- SMEs perceive a need for the help a technopole could provide in South Wales
- despite a shortage of R&D labs, other innovation support organisations provide a base for support in South Wales
- the services already available for SMEs in the region could be more widely diffused through a network
- South Wales has the appropriate regional coherence and identity
- UK Government policy had shifted towards support for initiatives of the kind a technopole represented.

The Pilot Project forum was particularly useful in building consensus between users and providers of innovation services in the region and the development of synergies between the participants. An important outcome of these developments was the forum's proposal that the members would take responsibility, at this stage without funding, for the prototyping of an electronic information network based on the *Internet* and *World-Wide-Web*. The forum believed that the development of an electronic network, as part of the technopole project, was important as it would facilitate South Wales-based firms to integrate at the regional level and also allow them to present themselves on a global level. Therefore, the South Wales Technopole would give firms a world-wide presence through a regionally mobilised network, a factor which is coherent with recent UK and European competitiveness policy.

South Wales and European Competitiveness Policy

Traditional regional policy appears to have had little effect on the generally low innovation levels in South Wales compared with the more prosperous regions of Europe (Cooke and Morgan 1992). Despite its relative manufacturing strength, boosted by substantial inward investment during the 1980s, a major question-mark still exists regarding the innovativeness of firms in Wales. Without improvement in their product and process technologies, few will be able to meet the supply requirements of the advanced manufacturers presently located in South Wales or elsewhere (Price *et al.* 1994). Although the South Wales economy has witnessed a period of relative success since the mid-1980s, it will not be possible to consolidate growth and revival unless industry becomes more competitive, efficient and innovatory. A recent European Commission White Paper on Competitiveness (CEC 1994) states that combinations of technological innovations (information technology, biotechnology, new materials, etc.) have led to intangible investment (in research, patents, training, etc.) growing faster than capital investment. The White Paper argues that such intangible investment should be further promoted – particularly through policies to improve vocational training and the general support of human resources.

The promotion of intangible investment and the development of industrial co-operation are key factors for the further success of the South Wales economy. The White Paper also states that research policy should take fuller account of the needs of the market, notably by means of closer co-operation with the operators concerned. The objective of the feasibility stage of the South Wales Technopole project was to define the market and the services and it is envisaged that the focus of technopole would promote developments with regard to infrastructure and environment that have much in common with EC recommendations to increase industrial competitiveness, that is:

- to stimulate the spirit of enterprise
- to raise training standards
- to manage industrial change
- to take up new technologies

- to speed up dissemination of innovations
- to allow the reorganisation needed to boost industrial efficiency and to increase the value added by production.

The fact that wealth creation depends increasingly on intangibles is making the world economy more fluid and volatile. As a result, the mobility of factors of production and the capacity to combine them effectively has become more important in the current context of world-wide competition between business and nations (Porter 1990; Reich 1991; Thurow 1993). The parochial thinking on which most policies with a bearing on industrial competitiveness are based is gradually making way for thinking in the South Wales economy and industry to be in terms of world-wide networks. Certain industries and firms in South Wales are developing strategies which will enable them to anticipate and manage change more effectively (Cooke 1994; Cooke and Morgan 1994). The most successful companies have already adopted such a strategy – particularly through the development of integrated supplier networks (Huggins 1994), realising that the challenge is no longer UK or European but world-wide, putting increasing pressures upon businesses to remain innovative and commercially viable. Greater innovation can be achieved if transparent information and stronger competition allow wider dissemination of the progress made amongst customers and suppliers via the development of networks. Furthermore, awareness of current trends is particularly important to enable firms to adapt to market changes.

A previous survey, undertaken in 1991/2, into innovative activity in Wales showed the following results (Cooke and Morgan 1992):

- Gross expenditure on R&D as a percentage of GDP (1991) in Wales was 1.1% – lower than the UK (2.2%), less than half the rate for Germany and Japan (2.9%), but comparable with Italy (1.4%) and Spain (0.6%).
- In Wales, most innovation synergies are between firms rather than between firms and universities. SMEs are more active than the UK average in winning EU and UK innovation project funding.
- There is increasing demand from both large and smaller firms for intermediary centres that can provide services in technology assessment, information and applications.

In response to this situation, the Welsh Development Agency has instigated numerous initiatives aimed to benefit and promote local firms and to provide industries, in particular SMEs, with a supportive and efficient innovation infrastructure. So far, many of these activities have been undertaken on a stand-alone basis. The obvious next step is to link these initiatives together in a network which will enable firms to be put in immediate contact with the precise business support services they require. Support structures should also facilitate the development of partnerships between big businesses and SMEs and the establishment of networks and clusters within industrial sectors. Despite limited resources, Wales' innovative strength has come primarily from its dominant SME sector rather than its branch plants. The leading R&D sectors in Wales, according to Cooke and Morgan's 1991/2 survey, are biological, environmental, information technology and materials technology.

Since the early 1990s, SMEs have been the only source of employment growth in the Organisation for Economic Cooperation and Development (OECD) area, with EU research indicating a growth of around 25% or 3.6 million extra SMEs in the European Union as a whole during the 1980s. To take account of this effect, regional, national and European Union policies and support for SMEs – particularly in the field of innovation – have taken a greater prominence. SMEs in Wales have, on average, taken far more advantage of such support than their counterparts elsewhere in the UK. The Welsh share of UK SMART (SME Merit Awards in Research & Technology) awards ranged between 9% and 14% from 1989 to 1993, while Welsh SMEs secured 7% of the UK share of EU Third Framework funding between 1987 and 1991 (Cooke 1994). Such figures should be reflected against the fact that Wales only accounts for 5% of the total UK population and 4.6% of GDP.

Inward Investment

The traditional industries of coal and steel in South Wales have declined to the point where they no longer play a significant role in the regional economy. The drive to replace these industries with new, often foreign-owned, investment in the manufacturing and service sectors has been extremely successful compared with many other regions in the UK. The question this raises is how South Wales can further capitalise on this success. In the context of the South Wales Technopole, the fact that the decision-making processes of many firms are increasingly being devolved to regional management structures means there are some real opportunities for branch plants in South Wales to play a more innovative role in the regional economy. This will, however, depend on the region being clearly able and willing to exploit these opportunities (Price *et al.* 1994). Thus there is more scope for innovative activity between the branch plant and its local milieu, such as interactions with training colleges, suppliers and local and regional development bodies. The impact of inward investment on high-technology employment is an important factor in the promotion of innovation and, as Table 6.1 indicates, Wales has performed extraordinarily well in attracting more than 10% of foreign direct investment (FDI) in high-technology jobs in the UK between 1984 and 1992.

Whilst the IBB data may not be totally accurate, possibly suffering from under-recording, omission or over-optimism, it nevertheless highlights inward investment as one of the key mechanisms for establishing growth in high-technology industry in South Wales.

Table 6.1 The destination regions of FDI jobs within high-technology industry in the UK, 1984–92

	Manufacturing	*Services*
South-East	35281	3017
East-Anglia	1243	320
South-West	3979	3
West Midlands	12802	2539
East Anglia	2239	80
Yorks & Humbs.	4439	103
North-West	13162	108
North-East	9097	24
Wales	16833	424
Scotland	35209	375
Northern Ireland	13795	554
Total	**148079**	**7547**

Source: Invest in Britain Bureau

The Feasibility Study

The South Wales Technopole project covers the whole of Industrial South Wales – comprising the counties of Gwent, Mid-Glamorgan, South Glamorgan and West Glamorgan and the eastern industrial corner of Dyfed. The objective of the project is to establish a truly networked region as a new generation of technopole, acting as a focus for innovation-boosting strategic linkages between support services, and to establish a long-term plan for the future development of the region. The Centre for Advanced Studies, within the University of Wales College at Cardiff, was commissioned by the Welsh Development Agency to undertake the feasibility study for the project. The first part of the study (which is to be followed by the formulation of a strategic plan) has three major components:

- an *Infrastructure Audit*, which reviewed the region's existing assets, such as its research capacity, its science and technology parks, its sources of finance and its technology support institutions.
- a *Customer Audit*, which surveyed the market for the services the technopole could provide. Customers include start-up firms, large enterprises and SMEs in the following sectors: electronic engineering, IT and telecommunications, automotive and aerospace engineering, biotechnology, medical technology, environmental technologies, chemicals and packaging.

- an *Innovation System Analysis*, which analyses the nature and extent of interaction between companies, and between companies and both private and public service providers. A further aspect of this stage of the study is a comparison of South Wales' Innovation System with two other European regions – Denmark and North-Rhine Westphalia (Germany).

A key feature of the initial audit phase of the Technopole feasibility study was a questionnaire survey of 200 companies in the South Wales region. The sample achieved a balance between firms of different sizes, as well as incorporating firms from a wide range of sectors where there was known to be some degree of high-tech activity in Wales. The survey generated 81 responses achieving a response rate of 40.5%. The survey achieved a satisfactory response within the size of firm bands, with the spread of responses closely reflecting the original sample. The research was designed around a questionnaire consisting of six main sections as follows: general information; R&D/Innovation; market focus; supplying new technologies; sourcing new technologies and innovation support. The following sections outline the main results obtained from the survey.

Finance for Innovation

The average turnover of respondents increased by 19% (from £18,665,500 to £22,180,255) between 1990/91 and 1993/94. The manufacturing sector in South Wales has shifted away from traditional staple industries to more modern sectors where technological and scientific skills are relatively more important. This is reflected in the average R&D budgets of firms, which have increased by a significant 31% from £362,546 to £476,225. These factors indicate a positive association between growth performance and the undertaking of R&D. Significantly, the most prolific growth of R&D budgets was amongst small firms where budgets increased by 37% between 1990/91 and 1993/94.

Technical Activities

The most common form of technical activity was development work, which involved three-quarters of all companies, while 66% claimed to have undertaken some form or facet of the innovation process. At least half the companies in the survey undertook adaptation, invention, testing or measuring. Whilst 46% undertook analysis and 30% were involved in certification, these activities were less prevalent among small firms. Although best practice companies in South Wales do see innovation as an intrinsic part of their business strategy, within a philosophy of continuous improvement and total quality management, there are still a large proportion of less successful firms who could improve their position significantly through the upgrading of their technological activities. Such upgrading should give careful consideration to a balance between technology push and market pull if implementation is to show true competitive advantage. As the UK government's Working Group on Innovation (1992) final report rightly stated, to be considered successful, innovation of any kind must result in improved profit and competitiveness or 'profitable change'.

Innovation Constraints

Such is the recognised importance of new technology, both in raising efficiency and improving competitiveness in the market, it is of critical importance to identify those factors that constrain firms from innovating and introducing new technologies. Despite changing government and banking policies, the most serious constraint on innovation performance was funding and the cost of finance. Firms were also particularly constrained by a lack of management time and recruitment difficulties. Financial constraints are especially marked with regard to small firms – many of whom are relatively new and often of a fast-growth nature. Approximately 20% of companies reported a lack of information on market potential/volumes, technical difficulties and competitive position. Also, a little under 20% of companies cited a shortfall of information on marketing opportunities and sources of external know-how, as well as a lack of autonomy and difficulties in retaining staff and accessing consultants.

The survey revealed that lack of funding and finance are the major constraints on innovation. Clearly this is an issue which warrants further investigation. A number of studies have shown that a bank overdraft is still the main method of raising finance for most firms – particularly SMEs who are often regarded as high risk and, therefore, are subject to higher interest rates than large firms – rather than investment from the business community. Business angels – who have operated with tremendous success, for both themselves and the businesses in whom they invest, in the USA – are still a relatively untapped market in the UK. A report on small- and medium-sized manufacturing enterprises (Levy 1993) stated that the majority of business angels in the UK learned of new business opportunities through friends, business associates and the press. Accountants, solicitors and banks rarely put them in touch with 'capital hungry' business.

Sources of Information and Knowledge for Innovation

The capacity to utilise external sources of technical information to supplement and support in-house technological development of new products and processes is known to be an important attribute of many firms, particularly in the high-technology sector. This survey has confirmed the necessity for firms to have access to the global technological base for new ideas, and the growing salience of knowledge and information, as the key conditions for competitive advantage in what has been termed the post-mass production or 'information' economy (Amin and Thrift 1995).

Substantial use is made of suppliers and customers at a local, national and global level with over 40% of companies using technical information acquired from customer or supplier firms. The utilisation of 'knowledge-links' with academic institutions and industrial associations was also significant with 31% of firms using such associations and higher education institutes, as well as further education colleges. Importantly, only 11% of firms used government agencies or consultants.

Whilst most firms acknowledged that technical information was available to varying degrees, it was felt that possession of this information is often fragmented, making it both time-consuming and expensive to obtain. This is of particular importance to SMEs in South Wales – who are crucially dependent on information provision through external sources such as public sector research institutions, universities and larger firms. On average, SMEs in South Wales have lower outputs per employee than larger firms so there appears to be a distinct need for action to improve their competitiveness through improved information and knowledge sourcing.

Technology Support Services

Overall, the survey respondents perceived there to be no co-ordinated approach to their requirements regarding technological and innovation support services. Despite these factors, a large number of firms spoke highly of services associated with both the Welsh Office and the Welsh Development Agency. The technology support services of these 'public sector' bodies were drawn on fairly extensively by companies. Over 50% had used some form of technology support service, with 27% using the Welsh Development Agency and 19% the Welsh Office.

Around 5% of companies used the Patent Office, which is based at Newport in Gwent, and the Wales Medical Technology Forum. Less than 3% had used the Wales Value Relay Centre or industrial associations, whilst fewer still used the Training & Enterprise Councils (TECs) or the Wales Quality Centre. The Welsh Development Agency dominates the overall breakdown in technology support services, accounting for 41% of the total usage of support services, followed by the Welsh Office with 27%.

The large majority of firms still considered there to be a gap in the support available to industry with regard to innovation and the lack of overall responsibility for facilitating the access and dissemination of knowledge. Such firms would undoubtedly benefit – many expressing a positive interest – from the introduction of a local forum to discuss innovation, leading to a cross-flow of ideas and new opportunities, co-ordinated by an independent body. The survey provides convincing evidence that existing arrangements are not as productive as they should be, indicating that significant benefit could be obtained through the development of an effective regional innovation network.

External R&D Facilities

Despite the commonly held view that industry regards the commercial awareness of academics with a degree of scepticism, over 30% of companies had used the technical services of higher education institutes or further education colleges. It is significant to note that the majority of links with universities and higher education institutes are at a local or regional level, rather than further afield. This is undoubtedly due to a number of factors, one of which is that lack of resources means that many firms are over-dependent on their own contacts and information. This ultimately results in many firms being unaware of the wider range of R&D services available or any complementary R&D which has been undertaken elsewhere. One of the major problems for firms, particularly SMEs, is identifying the exact type of R&D they require, which could result in the wrong choice of establishment and a loss of time and money. Also, firms do not have any great awareness of what R&D facilities are available in educational establishments and other organisations in their own locality. If these factors are successfully addressed, many companies will be able to take advantage of R&D facilities which are unavailable in-house.

Markets for Technical Services

The survey shows that the largest group of collaborative partners, with regard to the provision of technical services, were firms from the same value-chain (Porter 1990) – particularly product development for suppliers and customers. Such vertical linkages were evident not only between firms in South Wales and the rest of the UK but also with firms in Europe. The provision of technical services for public organisations within the UK were significant, but, perhaps surprisingly, provision for Welsh public sector organisations was negligible.

Technical innovation work for other companies was undertaken by 37% of firms and, with regard to linkages outside of the region, 33% of firms had at some point provided services for firms in the UK and 26% for firms in Europe. A surprisingly high 20% of firms undertook work for UK government bodies outside Wales. However, less than 3% of firms had worked for governments in Europe or other parts of the world and less than 2% for government bodies in Wales. These figures indicate the need to establish networks of businesses from the European Union and non-member countries and also the active promotion of the instruments already established by the EU, such as the European Community Investment Partners (ECIP) scheme.

Networking and Improved Support

The main areas of service improvement required by firms and which could be accessed through a regional innovation network – each cited by around 40% of firms – were:

- partner-finding for R&D
- information on technology developments
- partner-finding to obtain finance
- assistance with product design/development.

Although firms undertake networking arrangements for a wide range of reasons, a report by the Small Business Research Centre (1992) suggests that the three most important are: to help expand the range of expertise or products, to assist in the development of specialist services and products required by customers and to provide access to new markets. The process of networking, therefore, provides distinct advantages, not least allowing firms to exploit economies of scale and scope. The Technopole survey supports these assertions with more than 30% of responding companies needing partner-finding services in order to access new markets and technical consultancy work. Respondents also felt that more importance should be placed on the need to select the right partner to assist in the development of the innovation. There was a general view that some form of intermediary is required to effect the development of such networks, that is, individual firms should not be solely responsible for organising collaborative ventures since this has resulted in many firms in South Wales having little knowledge of whether potential partners had the degree of understanding and awareness which they were seeking. Problems were also encountered regarding the amount of time and resources needed to identify and secure potential partners, particularly in the SME sector.

In the first part of this chapter it was argued that the formation of networks with other firms and organisations is a mechanism by which companies can extend their knowledge base and strengthen their market position. As the survey has illustrated, firms in South Wales do make frequent use of external sources of technical information and knowledge in developing new products or production processes. However, a large majority of companies expressed the need for better information and access concerning at least one of the following (each cited by 20–30% of all companies):

- existing technologies
- databases with details of technical services
- technology assessments
- business services such as marketing
- access to laboratory facilities
- introductions to private investors
- information on IPRs
- contract research
- access to hardware and software testing.

Whilst existing organisations do attempt to bridge these information gaps, the current lack of co-ordination and interfacing between intermediaries and industry means they are not adequately fulfilling this role.

Lessons from North Rhine-Westphalia and Denmark

In addition to the survey of firms in South Wales, an integral component of the feasibility study was the comparative analysis of best practice regional networking in North Rhine-Westphalia (NRW), Germany and Denmark. These case-studies appear appropriate, since, as the theoretical issues argued earlier indicate, greater innovation is achieved through the transparency of information and knowledge. Therefore, network forms of infrastructure, such as those deployed in NRW and Denmark, are highly appropriate examples of the creation of competitive advantage for firms. This is particularly true for SMEs in enabling them to collaborate in order to compete. This enables them to use design, technological upgrading, product innovation, financing and marketing services they would otherwise be unable to afford (Cooke 1994). In both NRW and Denmark extensive innovation-oriented networks embracing a wide range of organisations have been built up. These networks have facilitated the development of a knowledge infrastructure which is significant in terms of the supply of highly qualified employees and the stimulation of business start-ups through spin-offs. Furthermore, such networks stimulate technology transfer – linking together firms and other organisations to exchange information, upgrade skills and develop new trading relations.

Within NRW and Denmark most networks exist as decentralised forms, either within cities, municipalities or counties. These are seen as the most effective way of raising innovative and productive capacity. Moreover, regional action is perceived as more cost-effective in that networking arrangements maximise resources which are otherwise wasted on futile competition and also enable smaller localities to benefit from the resources of more highly-populated areas.

Since the 1980s an extensive technology transfer network has been built up in NRW, with universities and polytechnics setting up technology transfer offices. These offices offer a whole range of services to SMEs in the region and have been a key feature in the development of the knowledge infrastructure. They are considered an important tool, supporting the regional economy and accelerating its restructuring (Hassink 1992). Rather than imposing too high a level of institutional and hierarchical co-ordination on its technological infrastructure, NRW appears to have accepted that, at its present stage of development, overlapping technological networks are the most efficient way of promoting innovative activity. The state of NRW has adopted an interventionist approach by creating a public infrastructure of technology institutions which provide a range of services for industry, particularly via intermediary bodies which serve as linkage-points from universities and research centres to SMEs. Intermediaries act as the gateways of the knowledge network between firms and NRW's 50 higher education institutes, 40 An-Institutes (privately-owned application-oriented research centres), 31 research and development institutes and also the ZENIT organisation – NRW's Centre of Innovation and Technology. Together these organisations form a dense network of public and private science-oriented organisations (Huggins and Thomalla 1995).

The transformation of the NRW infrastructure into an effective network support system has not been determined by the mere number of institutions providing innovative services. It is when these institutions are linked together that their individual information supplies and problem-solving capacities add up to a support infrastructure with a high level of connectivity (Grabher 1993). The formation of new regional institutions and the thickening of the regional infrastructure have cast some doubts on its efficiency but one of the key advantages of these 'organic' networks is their ability to disseminate and interpret new information. (Grabher 1991; 1993) views NRW as consisting of loosely coupled networks creating opportunities for sharing the learning experience of co-operating partners that results from their exchange relations with third parties. The loose coupling of institutions thus increases the learning capacity of networks.

In general, NRW's technology policy, which has been followed consistently for several years, is now reaping its rewards with more than 800,000 workers employed in technology sectors such as robotics, microelectronics, laser technology, opto-electronics, biotechnology and environmental science. Networking has been a very effective tool for developing small R&D-based firms and turnover figures for start-ups in the technology centres are impressive. Therefore, despite there being a wide array of agencies, initiatives and institutions, the network structure gives it a transparency allowing easy access.

In Denmark there has been a focus on small, mainly local government, programmes aimed at stimulating firms to enter small inter-firm networks to encourage growth and innovation. These networks, or 'growth groups' as they are often called, consist of 10 to 15 firms working on a particular theme, for example, product development, marketing, management systems, etc. The network initially entails firms meeting for seminars over a period of 18 months, whilst also receiving individual help from a specialist consultant who acts as the network's *animateur* (Sweeney 1987). The local council pays 50% of the expenses and the firms the other 50%, with a total average cost of approximately £100,000 per network. The networks give firms, who would normally have no relationship with each other, a chance to co-operate and create synergies leading to more long-term relationships between the participants. According to the networks' *animateurs*, many of the problems that the networks are able to address are similar to those encountered by firms responding to the South Wales survey, that is, a lack of information concerning:

- finance
- product development

- competitive position
- technical advice.

The advantages of such networks include:

- allowing several firms to share the cost of consultants, who assist the firms through a process of development – most usually in the fields of quality, productivity, human resource development or market extension
- the ability to access other resources
- enabling firms to share experiences and thereby learn from each other.

The networks also have an effect on the commitment of firms to the developments that the network is hoping to achieve. As Nielsen (1992) has argued, whereas agreements internally, or with a consultant, can be ignored if other business is pressing, an agreement among 10–15 firms to undertake some defined activities before the next meeting of a network is far more binding. Neither the manager nor owner of any of the firms involved would want to be the one to admit in front of his peers that he was unable to carry out the agreed work.

As well as the 'growth group' concept, another important development in Denmark has been initiated by the Danish Technological Institute (DTI) which has introduced a technology transfer network called Technology Partnerships. In the Technology Partnerships network the Institute is perceived to be a company facilitating access to technical knowledge, so that it becomes part of the complexes and chains as a specialised supplier, partner and customer. The DTI believe that the success of the technological infrastructure of Denmark depends heavily on its ability to deliver global knowledge in forms that generate here-and-now results, since the time horizon for companies before they expect first pay-back is rarely longer than one year. The DTI also realises that such global knowledge needs to be delivered to SMEs through local or regional networks (Nielsen 1994). The DTI is constantly increasing its access to global knowledge networks, involving contract technological research staff throughout Europe, the USA and Japan, through which it is able to circulate the innovation problems of Danish firms. Recent collaborative projects have linked Danish companies with research organisations attached to bodies such as NASA. The DTI envisages its future as acting as a normal company supplier with the goods provided being technological know-how or processing. A key feature of the DTI's Technology Partnerships network is that all companies who are members of the network are connected via an on-line electronic network. Therefore the network can make it possible for several individuals to work on a project, often without the need to spend time travelling to firms to give advice or assess a problem. However, the DTI does appreciate that electronics links alone are not enough and should be seen as complementary, rather than instead of, face-to-face 'human networking'.

Conclusion

It is clear that an important element of recent theory in the field of regional dynamics is based on network concepts: in terms of information, knowledge and innovation networks favouring the competitive advantage of regions. A capacity to network, which ties a region to relevant external partners, may therefore become a stronger determinant for development than many other previously important internal factors (Bergman *et al.* 1991; Nijkamp *et al.* 1994). The emergence of networks has enabled regions to be successfully integrated into global flows of information and knowledge, replacing the 'space of places' by a 'space of flows' (Castells 1989). If there is to be a serious opportunity to stimulate these flows, regions need to ensure that the accumulation of knowledge networks intended not only to produce knowledge-transfer but also a more general disposition to collaborate (Amin and Thrift 1995). Those regions which are successful in forging these links may be termed global regions and are likely to witness a significant increase in competitiveness and rapid economic development.

Although South Wales can by no means be viewed as a global region, its technopole project is seeking to establish the feasibility of the development of an innovation network based on best-practice networking strategies typified by Denmark and North Rhine-Westphalia, Germany. Such networks link together firms and other organisations to exchange information and knowledge, upgrade skills and develop new trading relations. The Technopole feasibility study has confirmed that the main advantage to be gained from a regional innovation network would be competitive advancement, creating more market opportunities and increased turnover and profitability through the bridging of information and knowledge gaps which currently exist within the sphere of innovation in the regional economy. The network would facilitate information to become more widely disseminated and transparent, allowing higher levels of innovation to be achieved. Although existing organisations, such as the Welsh Development Agency, Welsh Office and TECs, do attempt to bridge these gaps, the current lack of co-ordination and interfacing between intermediaries and industry means they are not fulfilling this role effectively.

The main types of services required by firms were partner-finding for R&D, information on technology developments, partner-finding to obtain finance and advice on product design and development. The primary users of the network will be SMEs who are already seeking partners, not only on a regional or national level but also on a European or world-wide basis. Therefore, the South Wales Technopole must not be constructed in isolation but be integrally linked to the increasingly prevalent system of world-wide networks, such as those currently available through the computer-based *Internet*.

The most serious constraint on innovation performance, particularly among smaller firms was funding and the cost of finance. The importance of cost factors in constraining the introduction of new technologies reinforces the view that SMEs are severely disadvantaged by the cost of innovating and the difficulties of raising funds for innovation. Lack of resources, including management time, has had the effect of making firms over-dependent on their own contacts and knowledge. This ultimately results in them not being aware of the wider range of R&D services available or any complementary R&D which has been undertaken elsewhere.

The majority of innovative firms in South Wales still consider there to be a gap in the support available to industry, indicating that firms would undoubtedly benefit from the introduction of a regional innovation network promoting the cross-flow of ideas and new opportunities whilst systematically streamlining procedures to make them more transparent. Such a network should take advantage of the latest developments in telecommunications, which is increasingly acting as a great leveller amongst SMEs and their larger counterparts. SMEs can now benefit from facilities available on public data networks. Research by Mercury Communications shows that small businesses are comparatively much higher users of telephones, faxes and mobile phones. The development of an innovation network in South Wales will result in there being very few facilities available to larger organisations from their private networks which will not be available to the larger pool of SMEs.

References

Amin, A. (1992) 'Big firms versus the region on the Single European Market.' In M. Dunford and G. Kafkalas (eds) *Cities and Regions in the New Europe: The Global-local Interplay and Spatial Development Strategies.* London: Belhaven.

Amin, A. and Thrift, N. (1994) 'Living in the global.' In A. Amin and N. Thrift (eds) *Globalization, Institutions, and Regional Development in Europe.* Oxford: Oxford University Press.

Amin, A. and Thrift, N. (1995) 'Institutional issues for the European regions: from markets and plans to socioeconomics and powers of association.' *Economy and Society 24,* 1, 41–66.

Bergman, E., Maier, G. and Tödtling, F. (1991) 'Reconsidering Regions.' In E. Bergman, G. Maier and F. Tödtling (eds) *Regions Reconsidered: Economic Networks, Innovation and Local Development in Industrialised Countries.* London: Mansell.

Camagni, R. (1991) (ed) *Innovation Networks: Spatial Perspectives.* London: Belhaven.

Castells, M. (1989) *The Informational City.* Oxford: Blackwell.

Castells, M. and Hall, P. (1994) *Technopoles of the World: The Making of Twenty-first-century Industrial Complexes.* London: Routledge.

Charles, D. and Howells, J. (1992) *Technology Transfer in Europe: Public and Private Networks.* London: Belhaven.

Christensen, P., Eskelinen, H., Forström, B., Lindmark, L. and Vatne, E. (1990) 'Firms in networks: concepts, spatial impacts and policy implications.' In S. Illeris and L. Jakobsen (eds) *Networks and Regional Development.* Copenhagen: University Press Copenhagen.

I would like to thank Julie Davies, Rob Wilson, Phil Cooke and Sara Davies for their help in producing this chapter.

Commission of the European Communities (1994) *An Industrial Competitiveness Policy for the European Union.* COM(94) 319 Final, Brussels: European Communities.

Cooke, P. (1994) *The Co-operative Advantage of Regions.* Paper presented to conference on 'Regions, Institutions and Technology: Reorganizing Economic Geography in Canada and the Anglo-American World', University of Toronto, September.

Cooke, P., Huggins, R. and Davies, S. (1995) *South Wales Technopole Project: Customer Survey and Design of Network, 2nd Interim Report to the Commission of the European Communities DGXIII SPRINT-RITTS Programme.* Cardiff: Centre for Advanced Studies.

Cooke, P. and Morgan, K. (1992) 'Regional innovation centres in Europe.' *Regional Industrial Research Report, No. 11.* University of Wales, Cardiff.

Cooke, P. and Morgan, K. (1994) 'The creative milieu: a regional perspective on innovation.' In M. Dodgson and R. Rothwell (eds) *The Handbook of Industrial Innovation.* Aldershot: Elgar.

De Bresson, C. and Amesse, F. (1991) 'Networks of innovators: a review and introduction to the issue.' *Research Policy 20,* 363–379.

de Vet, J. (1993) 'Globalisation and local and regional competitiveness.' *STI Review 13,* 89–122.

Gomes-Casseres, B. (1994) 'Group versus group: how alliance networks compete.' *Harvard Business Review,* July/August, 62–74.

Grabher, G. (1991) 'Rebuilding cathedrals in the desert: new patterns of co-operation between large and small firms in the coal, iron and steel complex of the German Ruhr area.' In E. Bergman, G. Maier and F. Tödtling (eds) *Regions Reconsidered: Economic Networks, Innovation and Local Development in Industrialised Countries.* London: Mansell.

Grabher, G. (1993) 'The weakness of strong ties: the lock-in of regional development in the Ruhr area.' In G. Grabher (ed) *The Embedded Firm: On the Socio-Economics of Industrial Networks.* London: Routledge.

Hassink, R. (1992) *Regional Innovation Policy: case studies from the Ruhr area, Baden-Wurttemberg and the North East of England.* Utrecht: NGS.

Hilpert, U. (1992) *Archipelago Europe – Islands of Innovation. Fast/Monitor FOP 242, Vol. 18.* Brussels: Commission of the European Communities.

Huggins, R. (1994) 'Local sourcing: machine tooling and inward investment in Wales.' *Welsh Economic Review* 7, 2, 37–47.

Huggins, R. and Thomalla (1995) 'Promoting innovation through technology networks in north Rhine-Westphalia.' In P. Cooke (ed) *The Rise of the Rustbelt.* London: UCL Press.

Johansson, B. (1991) 'Economic networks and self-organization.' In E. Bergman, G. Maier and F. Tödtling (eds) *Regions Reconsidered: Economic Networks, Innovation and Local Development in Industrialised Countries.* London: Mansell.

Karlsson, C. (1994) 'From knowledge and technology networks to network technology.' In B. Johansson, C. Karlsson and L. Westin (eds) *Patterns of a Network Economy.* Berlin: Springer-Verlag.

Karlsson, C. and Westin, L. (1994) 'Patterns of a network economy – an introduction.' In B. Johansson, C. Karlsson and L. Westin (eds) *Patterns of a Network Economy.* Berlin: Springer-Verlag.

Kogut, B., Shan, W. and Walker, G. (1993) 'Knowledge in the network and the network as knowledge: the structuring of new industries.' In G. Grabher (ed) *The Embedded Firm: On the Socio-Economics of Industrial Networks.* London: Routledge.

Lakshmanan, T. and Okumura, M. (1995) The nature and evolution of knowledge networks in Japanese manufacturing. Papers in Regional Science. *The Journal of the RSAI 74*, 1, 63–86.

Levy, J. (1993) *Small and Medium Sized Manufacturing Enterprises: A Recipe for Success.* London: Institution of Electrical Engineers.

Maillat, D., Crevoisier, O. and Lecoq, B. (1994) 'Innovation networks and territorial dynamics: a tentative typology.' In B. Johansson, C. Karlsson and L. Westin (eds) *Patterns of a Network Economy.* Berlin: Springer-Verlag.

Marceau, J. (1994) 'Clusters, chains and complexes: three approaches to innovation with a public policy perspective.' In M. Dodgson and R. Rothwell (eds) *The Handbook of Industrial Innovation.* Aldershot: Elgar.

Morgan, K. (1994) *The Fallible Servant: Making Sense of the Welsh Development Agency.* Papers in Planning Research, 151, Department of City & Regional Planning, University of Wales, Cardiff.

Nielsen, N. (1992) *Respective Roles of Government and the Market: Catalyst or Service Provider?* Paper to conference on 'Interfirm Linkages and Cooperation Among SMEs', Aspen Institute, Colorado.

Nielsen, N. (1994) *The Concept of a Technological Service Infrastructure: Innovation and the Creation of Good Jobs.* Paper to OECD Conference on 'Employment and Growth in the Knowledge-Based Economy'.

Nijkamp, P., van Oirscoht, G. and Oosterman, A. (1994) 'Knowledge networks, science parks and regional development: An international comparative analysis of critical success factors.' In J.R. Cuadrado-Roura, P. Nijkamp and P. Salva (eds) *Moving Frontiers: Economic Restructuring, Regional Development and Emerging Networks.* Aldershot: Avebury.

Porter, M. (1990) *The Competitive Advantage of Nations.* London: Macmillan.

Preer, R. (1992) *The Emergence of Technopolis: Knowledge-Intensive Technologies and Regional Development.* New York: Praeger.

Price, A., Morgan, K. and Cooke, P. (1994) *The Welsh Renaissance: Inward Investment and Industrial Innovation.* Regional Industrial Research Report, No. 14, Centre for Advanced Studies, University of Wales, Cardiff.

Reich, R. (1991) *The Work of Nations: Preparing Ourselves for 21st-Century Capitalism.* London: Simon & Schuster.

SBRC (1992) *The State of British Enterprise: Growth, Innovation and Competitive Advantage in Small and Medium-sized Firms.* Small Business Research Centre, Department of Applied Economics, University of Cambridge.

Scott, A. and Storper, M. (1992) 'Regional development reconsidered.' In H. Ernste and V. Meier (eds) *Regional Development and Contemporary Industrial Response: Extending Flexible Specialisation.* London: Belhaven.

Sweeney, G. (1987) *Innovation, Entrepreneurs and Regional Development.* London: Frances Pinter.

Thurow, L. (1993) *Head to Head: The Coming Economic Battle Among Japan, Europe and America.* London: Nicholas Brealey.

Tödtling, F. (1994) 'The uneven landscape of innovation poles: local embeddedness and global networks.' In A. Amin and N. Thrift (eds) *Globalization, Institutions, and Regional Development in Europe.* Oxford: Oxford University Press.

Working Group on Innovation (1992) *Final Report.* London: Department of Trade and Industry.

CHAPTER 7

The Emerging Shape and Form of Innovation Networks and Institutions

Andy Pratt

Introduction

The aim of this chapter is to offer an account of the changing shape and form of innovation networks and institutions in developed industrial economies. To date, this debate has been dominated by discussions of grand transitions from Fordist mass-production to post-Fordist batch production. Associated with this has been a concern with the social and economic contexts that are perceived to be necessary or sufficient to support, or promote, economic development. Researchers have highlighted the role of the institutional and network structures within which firms operate. A further dimension of the debate has had an epistemological character implicating either macro-structures (regulation theory) or micro-actors (flexible specialisation) in explanations of the transition process.

This chapter attempts to cut across these debates in an unorthodox fashion that is sceptical of the explanatory power, and extent of the applicability, of grand transition theories. It accepts that networks and institutions have always been important in industrial development; it is their exact nature, form and effect that are in question. Specifically, this chapter argues that the contemporary discussion about networks and institutions ignores the question of power. In order to understand the consequences and effects of different forms of economic arrangements for various actors and collectives, analyses of the relations between power and institutions are clearly required.

The key points of the chapter emerge through a review of different accounts of organisational networks and institutions. Accounts that rely upon, on the one hand, top-down macro-scale arguments and, on the other, micro-level actions are contrasted. The objective here is to resist the attractions of either position. Crucial to this aim is the concern with power; the chapter avoids the twin pitfalls of using power in an absolute totalising manner or eliminating power from the analysis altogether. The argument pursued rests upon a formulation termed 'power/institutional'. This approach attempts to explore how social relations are made and re-made across different scales, as well as how different actors are implicated in one another's strategies. The effect is a power differential between the agents involved. It is important that such analyses have a means of linking individual actions, work organisation and institutional co-ordination without reductive recourse to any particular agent.

The chapter is substantially concerned with the way in which innovation is accounted for, and configured in, organisations and the policy responses that have emerged to promote it. The chapter re-sits an account of innovation as either a one-off event or one that is simply contained in a workshop or laboratory: in effect it is concerned with the configuration of the whole of society. The chapter begins with a consideration of individualistic accounts of innovation which draw upon the metaphors of chains and linkages. This is followed by a section concerned with explanations of innovation that draw upon notions of structural features of economies as mediated by institutions. The third section draws out the problems associated with fixed, or ostensive, characterisations of social relations that are embodied in the foregoing accounts of innovation. In their place, a performative concept of social relations is suggested via the concept of 'power/institutions'. It is argued that such an account offers both a more satisfactory view of social relations and one that is sensitive to considerations of power.

From Entrepreneurs to Science Parks

Economic historians, such as Kondratieff, are often cited as discovering economic cycles; others, such as Mensch, have associated these cycles with different technologies (steam power, chemical and oil, electronics) – notably, Schumpeter suggested the importance of innovation 'swarms' in down-swings (see Marshall 1987). This has led to policy debates in which commentators have been concerned to identify or encourage innovation. The question is, how does innovation occur and how – if at all – can it be encouraged?

Writers on innovation, if coming from a positivist and/or neo-classical economic perspective, begin with the firm and the product cycle (Vernon 1966) plotting the stages of product development. Subsequent work by Markusen (1985) has developed this model by focusing upon profitability rather than the volume of production or sales. However, it does not actually tell us much about why some firms bring products to market and others do not. It simply suggests that the market does or does not the policy implication points to the importance of correcting such market failure. The usual responses by policy makers who share this view is to attempt to generate more new firms in the hope that they will generate more new innovations or products, some of which will be successful.

Focus on the market leads to an exclusive concern with exchange relations. Hence the importance of technology transfer between organisations, between universities and industry, and between intermediate producers. The hope would seem to be that the firms developing these new products and processes will be able to exploit market advantage and increase production. This begs one further question: why is technology not already transferred?

The standard response is to invoke the notion of the 'innovation chain' – like the product life cycle – monitoring the product from its initial inception as an idea through prototype, production, marketing and consumption. Some critics of such a conception of the 'innovation chain' concept have drawn attention to the presence of 'feedback loops' in the 'system' (see discussion in Rosenberg 1976; 1982). However, the overarching notion remains: it is linkages – or their lack – that are conceptually to the fore, together with an implicit acceptance of the discreteness of the elements within the innovation process.

Policy responses here seek to engineer more technology transfer via linkages between pure and applied scientists, and between applied scientists and innovating firms. In Britain the practice has been rather lacklustre with a range of rather ineffectual and unco-ordinated technology transfer programmes (see Miliband 1990). Ironically, it has been a private sector initiative – the UK Science Park Association (UKSPA) – that has internalised an understanding of the notion of technology transfer and has tried to codify it in an attempt to facilitate implementation. For the UKSPA (1990), a science park is: a development that has formal operational links with a higher education or research institution; that is designed to encourage the formation and growth of knowledge-based businesses and other organisations normally resident on site; and, the management of transfer of technology and business skills to the organisations on site.

From Institutions to Learning Regions

An alternative starting point is with the work of Schumpeter (1943) who outlined a conception of the innovation process as one of creative destruction, a process which would constantly undermine the very basis of economic activity through the development of new products, technologies and sources of supply. In their discussion of Schumpeter's contributions to this question, Freeman *et al.* (1982) highlight a shift in Schumpeter's writing over time with regard to the main agent of innovation from the small firm entrepreneur to the routinised research and development department of the large firm. Dosi (1983; 1988) argues that innovation is an evolutionary process that is specific to particular sectors of industry and linked to particular technologies: the macro context of innovation. Dosi uses the term 'trajectory' to express the notion of continuous incremental innovation allowing firms to evolve within a technology. Using the idea of paradigms from the history of science, Dosi suggests that 'revolutions' are rare, but new innovations – for example, semi-conductors or bio-technology – simply open up new trajectories.

Dunford (1993) argues that these approaches of Dosi and Freeman and his colleagues, termed 'Neo-Schumpeterian', generally avoid the reductionism inherent in explanations that draw upon the work of Kondratieff or Schumpeter. In contrast, these arguments focus upon an inter-related set of institutional factors rather than singular spatial, organisational or technological ones. Thus technological changes may have widespread effects on economic life; these may entail, by consequence, substantial and radical reorganisation, re-skilling and re-tooling of production processes. Dosi calls such a process a 'techno-economic paradigm shift'.

Massey *et al.* (1992) have drawn upon Dosi's work on innovation to offer an account of why science park development in the UK has been ill-conceived and ineffective. Their argument begins with the observation that the dominant mode of industrial organisation in the twentieth century has been large-scale mass-production; this production process is characterised by a rigid technical division of labour that has facilitated a functional separation of elements of the production process (termed 'Taylorism'). They argue that, over time, this functional separation has been extended by firms to create spatial divisions in an attempt to exploit possibilities of lower labour costs in some localities. Specifically, science parks are a logical extension of this process; representing a finer technical division of labour amongst scientists and researchers.

Massey and her colleagues suggest that Taylorism reinforces a conception of innovation as a sequential, linear process. They consider science parks to be weakened by the fact that they are tied into a very hierarchical and fractured production *and* innovation process. Whilst Fordism has many benefits, flexibility and innovation are not amongst them. This argument suggests that science parks are a child of the Fordist organisation of production. This corresponds with Schumpeter's view (see above). Science parks are the hiving-off, the sub-contracting out, of research and development activities. The social overhead costs of research and development are thus effectively transferred from the individual firm to the promoting organisation (invariably the [local] state, or the state by proxy: a university).

This account, although sensitive to spatial variation, does rely heavily upon a structuralist view of the labour process in which technology and control are wielded unambiguously by capital, or the state. Other analyses have sought to avoid such problems by drawing upon meso-level concepts such as institutions. Dunford (1993, p.40) follows such a line of argument in a call for the consideration of technopoles in the light of the broader 'institutional conditions in which inventions and research and development occur' and the 'different institutional conditions in stimulating the diffusion of technologies'.

Discussion of 'institutional conditions' requires some clarification. Taken at face value, Dunford's article might simply be read as a call for an approach informed by institutional economics (c.f. Williamson 1975; 1985). Indeed, such an approach might fit conveniently with the discussion of post-Fordism and new industrial districts by the well-known institutional economists Piore and Sabel (1984).

Whilst the empirical validity, or otherwise, of the post-Fordist hypothesis has been subject to near saturation debate, the epistemological assumptions of the institutional economics that underpins it has not. Institutional economics attempts to respond to the under-socialised and atomistic concept of economic action inherent in neo-classical economics, stressing instead the importance of the analysis of social institutional and transactional factors (such as decision making, rationality and trust).

It is important here to note the critique of institutional economics (see Granovetter 1985; Powell 1990; Hodgson 1988; 1993). Granovetter, in particular, argues that institutional economics tends to under-socialise economic action, for which it compensates with an over-socialised notion of society; this is underpinned by a functionalism that implies that institutions arise as efficient solutions to economic problems. In the place of institutional economics, Granovetter stresses the importance of understanding the social embeddedness of economic action. Recently, this approach has informed work into the social context of industrial development (see Amin and Thrift 1994; Grabher 1993). In particular, writers such as Lorenz (1990; 1992) have stressed the importance of relations of trust, or what Storper (1993) more generally has referred to as 'untraded dependencies'.

A particular articulation of this idea has been suggested by Powell and Dimaggio (1991) which they term 'institutional thickness'. Amin and Thrift (1992; 1994) elaborate this as 'a simultaneous collectivisation and corporatisation of economic life'; but they go on to stress the importance of the consideration of the dynamic process of institutionalisation rather than the static presence of institutions. Kevin Morgan (1995) draws upon this line of argument and relates it back to the innovation process via the concept of the 'learning region'. Morgan draws upon the work of Lundvall (1994) in order to highlight the role of knowledge and learning in economic development. Morgan's substantive concern is with the role of the Welsh Development Agency (WDA) in regional economic development. He charts the development of the WDA from a glorified property developer to an agency supportive of aftercare, supplier of development technology support and skills formation. The WDA has, for example, used branch plant firms as 'tutors' in the creation of business support programmes. Morgan stresses the interactive process of innovation and its contextual setting, which is shaped by a variety of institutional routines and social conventions. He then develops the concept of a learning region. The learning region is, in this context, a particular structured combination of institutions strategically focused on technological support, learning and economic development that may be able to embed branch plants in the regional economy, and hence cause firms to upgrade *in situ* rather than to relocate away from the region.

From Networks and Institutions to 'Power/Institutions'

How practical is Kevin Morgan's argument? Can the promotion of networks, or the existence of an institutionally rich locale, account for the form of economic development? It will be clear from the above that considerable emphasis has been switched from the individual entrepreneur to the individual firm in an institutional context. The flow of knowledge and ideas now takes place in two, or more, directions: Morgan (after Schoenberger 1994) suggests the possibility of such an institutional embeddedness changing the organisational position of the branch plant from a subservient one to an active innovator. In a sense, the institutional context is about reversing the flow of skill and control out of the region and creating an embedded economic structure. There is a key weakness: the analysis of power that underpins this account. This section focuses upon the character of the conceptual problems inherent in conventional network and institutionalist analyses and develops an alternative approach that is more sensitive to power.

The recognition of the weaknesses inherent in institutional approaches to the study of contemporary organisations has led a group of writers to propose an alternative: a 'power/institutions' approach (see Clegg 1990; Clegg and Wilson 1991). They argue that materialist approaches link power and technology, but their characterisation of technology is one in which technology conceals power, and hence protects capital's interests. On the other hand, 'engineering' views of technology conceptualise it as a neutral agent. From the former view, power is all pervasive, from the latter it is non-existent. Clegg's (1989) alternative is to draw upon post-structuralist arguments such as those developed by Foucault, in particular his work on disciplinary practice and governmentality (see Burchell *et al.* 1991). The concept of disciplinary practice highlights the ways in which the practices of individuals and organisations are defined and regulated by particular modes of rationality. In this context, a mode of rationality may correspond to scientific management as embodied in Tayloristic work organisation.

The very concept of Taylorism carries with it the notion of delimitation and control of a process. Within a defined process, 'jobs' are created as the smallest possible unit of process activity – moreover these tasks have to be co-ordinated: hence the need for 'managers' (see du Gay 1994). Workers, managers and the organisation are all governed via this mode of rationality – different decisions are justified or rejected according to its code (see Rose 1989). Most obviously, the modes of rationality of managers and their identities are created through the discourse of 'management science' (see Hoskin 1995). It is important to note several points here: first, that power is conceived as a *practice* rather than a *position* within a bureaucracy; second, that there is no *a priori* presumption of the direction, or mode of exercise, of power; third, institutions and structures do not precede agents, and agents do not arrive pre-formed either. In short, power is an effect, or an outcome; as such it cannot be ascribed in the abstract or generalised, it will be particular to, and occasioned by, specific situations.

A similar argument to that developed from Foucault's later work on governmentality has developed through research into the nature of scientific work (known as the sociology of scientific knowledge). A highly influential and clearly argued exposition of the process of innovation, and the relationship between science and technology from such a perspective, is provided by Bruno Latour (1987). Latour argues in favour of two key principles concerning the analysis of scientific work: first, the imperative of not focusing on the final products of science and technology but rather to follow them through the process of their creation. This study of *science and technology in the making* leads us, he argues, back from cold, stable objects towards warm, unstable objects. Thus, Latour argues that the process of science is centrally one of getting others to corroborate one's own findings to make unstable facts stable. To make them stable these findings must be reproduced or disseminated by others. This leads to Latour's second point that the process of the reproduction of ideas and findings is not a simple copying process but rather one of translation and transformation.

An analogy can be made with the parlour game of 'Chinese whispers' whereby a story is passed around a circle via whispers. The cumulative mishearing and misinterpretation produces a new story by the time it returns to its origin. However, in contra-distinction to chance mishearing in the game, Latour views transformations as active and purposeful. Moreover, the circle of people – the network – does not pre-exist and has to be created. So, the interests of a group of people at a dinner party have to be co-opted and co-ordinated before the game can be played in the first place. To be effective, the initiator has to enroll the interests of her guests into her plan; she has to make the guests believe that their objectives (having a good time) can only be achieved by playing this game. If they agree, they will submit to the rules and play. They will only continue to play as long as they believe their interests to be satisfied in this way. For example, the game might be disrupted by a guest encouraging others to stop playing Chinese whispers and to begin a game of charades.

This last point is echoed rather more formally by Clegg and Wilson (1991, p.243) who argue that 'agencies interested in maximising their strategicality must attempt to transform their point of connection with some other agency or agencies into a necessary "nodal point": a channel through which traffic between them occurs on terms which privilege the putative strategic agency.' In Latour's work, a significant 'nodal point' is the laboratory. Laboratories are, for Latour, the places where warm, unstable science is made stable and cold technology: after which it can be let out into the world to be used. After the title of an earlier paper (Latour 1983): 'give me a laboratory and I will raise the world'. But, the process of translation and fixing is not final, it does not stop at the laboratory door or the factory gate, different agencies and users will be constantly in the game of translating the technology into their own interests – this may again create further transformations.

Miller and O'Leary (1995) have drawn together these two strands of argument, from Foucault and the sociology of scientific knowledge, and used them to account for the transformation of work practices in a factory producing Caterpillar tractors. They see the factory, like Latour did the laboratory, as a crucial site of transforming and fixing technologies via instruments, ideas and calculations. In many ways the idea that factories are about production and construction is easy for us to grasp. However, they argue that factories are just as much about the construction of social practices as they are about constructing technical practices and products. In their study they stress how the introduction of new work practices – which would be commonly considered under the label of flexible specialisation – should be viewed as an attempt to create a new modality, one which is concerned with the construction of a new form of identity, or economic citizenship, as well as new factory layouts, production arrangements and products. They term this new modality, this new entity, 'an assemblage'. They stress that, as such, it is always fragile or unstable, as the outside world keeps 'intruding'. As soon as one element in this complex of interrelated practices and locales was altered or removed, then there was a possibility that the assemblage itself would be modified or transformed' (Miller and O'Leary 1995, p.32).

Innovation Networks and Institution Building: Making Science and Technology More Durable

I think that we can now reconsider Morgan's discussion of the learning region. Drawing upon the 'power/institutions' approach outlined above, we should become cautious of two elements of his story: first, the apparent lack of acknowledgement of power and second, the focus on meso-level bodies such as institutions. A 'power/institutions' account would highlight the role of inward-investing companies and their attempts to create a new modality of economic organisation, to stabilise their production techniques and technologies, and their workers. Such a modality could be created in a variety of contexts; at the beginning it is simply the idea of a market – a factory. The practical establishment of a factory involves a huge range of complex tasks that imply co-ordination as well as a particular definition of the nature of work and management. The managers of the firm seek to enrol the potential workers, the local sub-contractors and existing government agencies into their particular task. Aside from the necessity of convincing all of the agents that they should want to be enrolled – perhaps not so difficult in this case – enrolment has to take place.

Agents will have their own interests; they will naturally also seek to achieve them at the same time, in fact the inward-investing firm's objectives will only be achieved if they can. In this context it is not surprising to find that the WDA – which clearly has an agenda – attempts to mobilise itself and others to capture the inward-investing firm to further its aims: regional development. At each turn the interests get transformed: agencies like the WDA were created and, we learn from

Morgan, actively transformed from property developer to 'animateur' in the process of attracting inward investment. Likewise, *pace* Miller and O'Leary, the system of production developed by the company if established in South Wales will be different to the one established at plants elsewhere in the world.

This brief comment on Morgan's work serves to highlight what sociologists term the 'performative' nature of social relations and the active nature of power. Different entities or assemblages (firms, development agencies, locales, regions, individuals) are constantly remaking themselves, attempting to enrol and being enrolled. Those agencies that succeed in getting others to perform for them – through delegated tasks – create for themselves most power; power that increasingly allows the coercion of others. Delegation, in this context, does not have to be seen as happening within an organisation – if a set of rules and norms are set by an agency and tasks denoted by such rules and norms are performed by others of their own accord, then clearly this is a subtle and effective exercise of power.

Innovation requires context, but context – as if a backdrop – is insufficient. Innovation is a practice that requires co-ordination. Entities created by innovation are unstable: they can only be created under heavily controlled conditions and, with much effort and expense, made stable under such conditions. Entities (products, networks or institutions) can only move out into the world if other agents and interests can be enrolled to reproduce the laboratory or factory in the world. To make any of these entities operative requires the building of networks; it also requires the exercise of power. Networks should not be seen as neutral, neither should they be seen as unproblematic conduits of power. Clegg and Wilson (1991, p.266) highlight this paradox, such that no power is absolute: there is always a space for discretion. 'What is crucial is the subordinates' preparedness to bend managerial control and management's success in securing the kind of power circuitry which enables them to do so'. They argue that this is not a zero-sum power relation, rather that positive sum outcomes can occur and when they do they may offer fruitful contingencies and create a place for real politics.

So, institutions or networks must be seen as constantly under construction, not ready made; they are unstable but can be made more stable. Moreover, they have a power effect, they are not neutral. The effect is the creation of an identity of a production-line worker or manager, the production of a new product, the organisation of a factory, or the configuration of a region. Effects cannot be determined *a priori*, rather they are outcomes of an ongoing progress of struggle to re-define agents' interests and co-operation. The building of locality-based networks is an attempt to enrol actors into that locality; in the words of institutionalists, to embed them. However, embeddedness is, at best, temporary, and it is never total. As scientists build their laboratories, managers build their factories, and regional development agencies build their regions: the trick is to make networks and alliances more stable.

We can also reflect upon the account of science park development outlined by Massey *et al.* (1992) and referred to earlier in this paper. In this account, institutions and technology have a role, but one that is subservient to capital/power. So, science parks became an extension of the spatial and technical division of labour. A re-reading from a power/institutionalist perspective might suggest that science parks were an example of a proto-agent, or an assemblage, that sought to mobilise resources to build new products. The network builders in this case were predominantly universities, property developers, local authorities and agents of central government. Science park promoters were most successful in getting property-related concerns to pass through their 'nodal point' and to collaborate with them. For local authorities the potential advantage was job creation and technology transfer to promote other firms. For property developers the interest was a return on investment. These agencies were less successful in getting firms to locate there who wanted to engage in technology transfer; many just wanted nice property in a good location with a prestigious address. Like the factory described by Miller and O'Leary (1995), the science parks were unstable assemblages.

Many science parks were unable to recruit sufficient firms or they recruited the wrong firms (non-innovators). There is evidence of the instability of science parks manifest in an identity crisis through the concern with potential confusion with business parks; this accounts for the attempts by park managers to create their own network (UKSPA) – central to which is a 'membership criterion' of certain shared characteristics of the science park (see UKSPA 1990). There are some further interesting dimensions that have emerged from Massey's (1995) work, which focuses upon the intersection of gender relations, identity and work culture on (and off) science parks and highlights the significance of the regulation of identity and masculinity of male workers in high-tech firms.

The arguments developed above are suggestive of the similarities between the development of science parks and the construction of institutions and networks. What is different is the effect. The science park builders were, perhaps, less ambitious in their attempts to enrol actors by physically limiting their domain of action: they sought to bring together all of their resources at one (physical as well as organisational) point. The network builders of South Wales were less limited in their spatial scope. They spun their webs more effectively by enrolling agents at all levels: from individuals, to firms, to organisations; they then sought to articulate them together in order to achieve their ends. This is not to suggest that there has been a unitary agent in control or that all objectives have been achieved. What is particularly interesting in the WDA case is, as Morgan notes, that the WDA has itself been changed, been reconfigured, in this process.

Conclusion

This chapter has stressed the importance of a concern with networks and institutions. However, it has been very critical of their conception. The 'power/institution' argument counters the idea of institution as 'context' or as a meso-level intermediary; it also resists the representation of networks as neutral or 'cloaks' of power. It does not presume that agents (which may be individuals, firms or regions) come pre-formed, or ready made. Its central concern is with the ways in which agents are constructed and translated in practice; it is also concerned with what the effects of such translations are in terms of power.

This chapter has suggested that the 'grand transition' debate can be side-stepped. What is important is the analysis of the processes by which stabilisation of objects and agents happens and what power effects they have (see also Murdoch 1995). This approach implies a rather different perspective on innovation, economic and technological change. It encourages researchers to follow policy makers, firms and products, through their networks into society. Agents that seek to achieve a position of stategicality within networks would do well to pay attention to the processes and techniques of network building and the translation of ideas and technologies. Building a network, a science park, or a better mouse trap, is not sufficient unless others can be convinced of its value and be enrolled into its promotion and use.

References

Amin, A. and Thrift, N. (1992) 'Neo-Marshallian nodes in global networks.' *International Journal of Urban and Regional Research 16*, 571–587.

Amin, A. and Thrift, N. (1994) (eds) *Globalization, Institutions and Regional Development in Europe. Oxford: Oxford University Press.*

Burchell, G., Gordon, C. and Miller, P. (eds) (1991) *The Foucault Effect: Studies in Governmentality.* Hemel Hempstead: Harvester/Wheatsheaf.

Clegg, S. (1989) *Frameworks of Power.* London: Sage.

Clegg, S. (1990) *Modern Organizations: Organizational Studies in the Post-Modern World.* London: Sage.

Clegg, S. and Wilson, F. (1991) 'Power, technology and flexibility in organizations.' In J. Law (ed) *A Sociology of Monsters: Essays on Power, Technology and Domination.* London: Routledge.

Dosi, G. (1983) 'Technological paradigms and technological trajectories.' In C. Freeman (ed) *Long Waves in the World Economy.* London: Buterworths.

Dosi, G. (1988) 'The nature of the innovation process.' In G. Dosi, C. Freeman, R. Nelson, G. Silverberg and L. Soete (eds) *Technological Change and Economic Theory.* London: Frances Pinter.

du Gay, P. (1994) *Making up Managers: Expertise and the Ethos of Bureaucracy.* Paper presented at the History of the Present meeting, LSE. Paper available from Dr P. du Gay, Department of Sociology, Open University, Walton Hall, Milton Keynes.

Dunford, M. (1993) 'Technopoles: research, innovation and skills in comparative perspective. In J. Simmie, J. Cohen and D. Hart (eds) *Technopole Planning in Britain, Ireland, and France.* Working paper 6, Planning and Development research centre, The Bartlett, UCL, London.

Freeman, C., Clark, J., and Soete, L. (1982) (eds) *Unemployment and Technical Innovation.* London: Pinter.

Grabher, G. (ed) (1993) *The Embedded Firm: On the Socioeconomics of Industrial Networks.* London: Routledge.

Granovetter, M. (1985) 'Economic action and social structure: the problem of embeddedness.' *American Journal of Sociology 91*, 481–510.

Hodgson, G. (1993) *Economics and Evolution: Bringing Life Back into Economics.* Cambridge: Polity.

Hodgson, G. (1988) *Economics and Institutions.* Cambridge: Polity.

Hoskin, K. (1995) *Recalculating the Firm: Appraising the Power of Powerful Knowledge.* Paper presented at the History of the Present meeting, LSE. Paper available from Dr K. Hoskin, Warwick Business School, University of Warwick, Coventry.

Latour, B. (1983) Give me a laboratory and I will raise the world. In K. Knorr-Certina and M. Mulkay (eds) *Science Observed: Perspectives on the Social Study of Science.* London: Sage.

Latour, B. (1987) *Science in Action.* Cambridge, Mass: Harvard University Press.

Lorenz, E. (1990) *The Social Construction of Trust: Informal Networks of Sub-contracting in French Industry.* Paper presented at Conference on Flexible Specialisation in Europe, Zurich.

Lorenz, E. (1992) 'Trust, community and cooperation: toward a theory of industrial districts.' In A. Scott and M. Storper (eds) *Pathways to Industrialisation.* London: Routledge.

Lundvall, B. (1994) *The Learning Economy: Challenges to Economic Theory and Policy.* Paper presented at the EAEPE conference, Copenhagen 27–29 October.

Markusen, A. (1985) *Profit Cycles; Oligopoly and Regional Development.* Cambridge, Mass: MIT Press.

Marshall, M. (1987) *Long Waves of Regional Development.* London: Macmillan.

Massey, D., Quintas, P. and Wield, D. (1992) *High Tech Fantasies.* London: Routledge.

Massey, D. (1995) 'Masculinity, dualisms and high-technology.' *Transactions of the Institute of British Geographers 20*, 4, 487–499.

Miliband, D. (1990) *Technology Transfer.* Institute of Public Policy Research, Industrial Paper No 2, IPPR, London.

Miller, P. and O'Leary, T. (1995) *The Factory as Laboratory.* Paper presented at the History of the Present meeting, LSE. Paper available from Dr P. Miller, Department of Accounting and finance, LSE, Hougton St, London.

Morgan, K. (1995) *The Learning Region: Institutions, Innovation and Regional Renewal.* Papers in planning research, No 157, Department of City and Regional Planning, Cardiff, University of Wales.

Murdoch, J. (1995) Actor-networks and the evolution of economic forms: combining description and explanation in theories of regulation, flexible specialization, and networks.' *Environment and Planning, A* 27, 731–757.

Piore, M. and Sable, C. (1984) *The Second Industrial Divide: Possibilities for Prosperity.* New York: Basic Books.

Powell, W. (1990) Neither market nor hierarchy: network forms of organization. *Research and Organizational Behaviour 12*, 74–96.

Powell, W. and Dimaggio, P. (ed) (1991) *The New Institutionalism in Organisational Analysis.* Chicago: Chicago University Press.

Rose, N. (1989) *Governing the Soul: the Shaping of the Private Self.* London: Routledge.

Rosenberg, N. (1976) *Perspectives on Technology.* Cambridge: Cambridge University Press.

Rosenberg, N. (1982) *Inside the Black Box: Technology and Economics.* Cambridge: Cambridge University Press.

Schoenberger, E. (1994) Corporate strategy and corporate strategists: power, identity and knowledge within the firm. *Environment and Planning, A* 26, 435–451.

Schumpeter, J. (1943) *Capitalism, Socialism and Democracy.* London: Routledge.

Storper, M. (1993) Regional worlds of production: learning and innovation in the technology districts of France, Italy and the US. *Regional Studies 27,* 433–455.

UKSPA (1990) *Summary of Operational Science Parks in the UK.* UKSPA, Aston Science Park, Birmingham.

Vernon, R. (1966) International investment and international trade in the product cycle. *Quarterly journal of economics 80,* 190–207.

Williamson, O. (1975) *Markets and Hierachies.* New York: Free Press.

Williamson, O. (1985) *The Economic Institutions of Capitalism.* New York: Free Press.

CHAPTER 8

The Japanese Technopolis Strategy

Sang-Chul Park

Introduction

Japan, along with West Germany, is regarded as the country of economic miracle. After the Second World War, the Japanese government tried to overcome its technological inferiority to the Western countries and, at the same time, practiced a high-productivity and growth-oriented economic policy.

As a result of this policy, Japan exported more of its goods to the European Community (European Union since December 1993) and to the United States of America during the 1980s than it imported from these countries. The Japanese export goods now dominate all of the world markets, making Japan the biggest financial country in the world. During the dramatic increase of its export products in the 1960s and 1970s, Japan was criticised for imitating technology originating in the Western countries. Therefore Japan started to develop its own technical capacity and its efforts were of national concern at that time. Technology parks were built in several places and a science city in Tsukuba. In spite of such efforts, Japanese technology parks were merely technology transfer centres from the Western countries, instead of developing their own creative technologies.

In order to solve the fundamental problem the Government, in the early 1980s created a new plan for technopolises as part of a research and development programme in which local regional authorities would be responsible for regional development.

The development of high-technology in Japan has been part of the economic policy and the central government, especially the MITI (Ministry of International Trade and Industry), has played a very important role in advising Japanese industries of the government's guidelines and industrial policies.

Purpose of the Study

Taking a medium- and long-term view, there appear to be some important questions remaining as to whether the emergent new high-technology industries (Microelectronic, Biotechnology, New energy technology and New material) will give a sufficient boost to regional development. Two important questions are: can the high-technology industries produce employment opportunities, in contrast to 'mature' heavy industry in the present depression and decline, and, can they serve to develop or reconstruct the less-developed, unemployed areas?

With these issues I shall attempt to explore how the technopolis plan is being carried out by MITI and the local governments, why they chose the new high-technology in the fields of the above implied areas and how the technopolis plan will affect Japanese political economics and regional development.

Methodology of the Study

For the methodology of the study I focus on an empirical level, and the theoretical level, which will contribute to understanding the technology policy in general. Certainly the former is likely to be stressed because a presentation of the technopolises will be more effective than any other method in understanding the regional situation and conditions. As regards the latter, theoretical backgrounds of the industrial policy based on the macro-economy and locational pattern of the technopolises with respect to economic geography will be discussed.

Validity and Reliability

It is too early to conclude the technopolis plan because it is a long-term-based strategy for the twenty-first century and still in the process of development. According to government statistics, 20 of the total 26 technopolises had achieved the first step in building up infrastructures such as airports, railroads, super-highways and research and development centres by April 1991. The second step began in 1991 to complete the project by the year 2000.

The technopolis plan does not affect the Japanese economy severely in the short-term, but, in the long-term, it will contribute not only to boosting the domestic economy but also to improving technological standards in the regions, and even to minimising technological inequality between the metropolitan areas of Tokyo, Osaka, Nagoya and the regions. This tendency has already taken place in the IC-Chips industry in the southern main island of Kyushu.

As hindrances of the technopolis plan, the financial dependence of local governments and the lack of highly qualified human resources in regions are noted. Both factors result in strong competition between the technopolises and keep Japanese and foreign firms apart.

The Conception of Technopolis

The technopolis plan was proposed by MITI in 1980 after Japan had discussed the future of its industrial policy for a long time. The proposal was first made in *The Version of MITI Policies in the 1980s*, a publication issued by MITI's Industrial Structure Council in March 1980. It became a definite government policy with the passage of the *Law for Accelerating Regional Development based upon High Technology Industrial Complexes in 1983* (Kodansha 1993). At present, there are altogether 26 technopolises in Japan. The conditions for the technopolis are as follows: a minimum population should be 200,000 in the technopolis area, it should be possible to make day trips to the three large metropolitan areas, either by air or by bullet train (Shinkansen).

The technopolis consists of three components: an industry zone, an academic zone, and a habitation zone. In the industry zone, the production facilities of high technology industry should be built. In the academic zone, various facilities for research and education, such as private and public R&D institutes and universities, are necessary. The habitation zone requires living facilities for managers, engineers, scholars and their families (Hahne 1985; Toda 1990). The most important structure of the technopolis was finished in the beginning of 1991 and it will be developed continuously in the 1990s.

Properly speaking, the conception of technopolis is a part of the industrial plan. Japan produced its entire industrial plan in the beginning of the 1980s in order to improve the competitiveness of industries. The industrial plan is composed of six different strategies: joint R&D projects, strategic alliances with foreign countries, technopolis plan, telecommunication network, venture capital and venture business, and promotion of selective imports. None of these six strategies are based on a large capital increase from the government or by its direct intervention in the reconstruction of industry. They are in agreement with maximum applications of industry in a better transformation of society and promote private investments in appropriate areas (Tatsuno 1986).

The word, 'technopolis', which denotes two different meanings, symbolises Japanese industrial strategies in the 1980s and 1990s. The first meaning, technology, aims at updating Japan's industrial structure to move the nation toward the goal of being a 'high-tech archipelago' in the twenty-first century. The second meaning, polis, stems from the Greek word for a city state and pursues the development in areas away from the leading industrial and cultural centres of Tokyo, Osaka and Nagoya by accelerating the transfer of technology to regional industries (Japan Industrial Location Center 1990a). Japanese regions attempt to hold a balance between the profit of private industries and the public interest. In order to realise both ideas, MITI has developed the technopolis plan continuously. It regulated the co-operation between the industry (San), scholars (Gaku) and local government (Kan). Nevertheless, MITI will not intervene in the technopolises. The local governments take their own initiative and try to improve their economic development further. The central government only supports efforts to reach regional deconcentration (Look Japan 1984). In addition, local governments should create the best conditions for research and production facilities. This results in strong competition between the technopolises.

Theoretical Approaches to the Locational Pattern of High Technology Industries and the Technopolis Plan

R&D, production arrangements and corporate organisations interact in a dynamic process. As a theoretical model, the theory of R. Vernon, the product-cycle model, can explain the dynamic locational process. Vernon argued that the development and arrangement of multi-nationals depends on the product cycle: new products, matured products and standardised products. According to him, an enterprise in the United States – where higher income, wages and technologies are prevalent – monopolises profit by high value-added new products on the domestic market and it expands its market to other industrialised countries, such as Japan and Western Europe. As a second step, imitators or competitors produce the same products in and outside the United States. Therefore, the price of products starts to fall. During this time, the enterprise starts capital investment directly in other developed countries to manufacture the products locally. In doing so it could share the markets in other developed countries on the one hand and, coincidently, re-import to the United States at lower prices on the other. This pattern will be extended further from other developed countries to developing countries (Vernon 1966).

With Vernon's model, the locational pattern of high-technology industries in Japan can be explained as well: in the early stage of the product cycle, an area where knowledge, information and technology are easily obtainable will be more advantageous – such as the three metropolitan areas of Tokyo, Osaka and Nagoya. As a second stage, the dependency on a technology foundation is decreased and the decentralisation of manufacturing from the nucleus area follows. In this case, the field of R&D will be decentralised first because it is technology-improvement and less dependent on manufacturing directly. This area forms satellite cities around the three metropolises and parts of Kyushu Island, such as Chiba, Saitama, Kanagawa, Aichi, Mie, Hyogo, Nara and Oita. At the final phase, locations for production moves towards areas with cheaper prices, because standardisation of products is complete and new technology investment is no longer necessary. These areas are located on the east coast and Kyushu Island (Nishioka 1989).

The locational requirements of high-technology industry are various. Accessibility to academic research institutes and living conditions – urban amenities, housing, and natural, traditional and cultural environment – are regarded as the most important factors. Surely the factors influencing the location of industrial activities, such as access to markets and infrastructure, may not be overlooked (Toda 1990)?

In relation to the technopolis plan, it is expected that high-tech industry from central areas transfer high-technology to smaller firms in the region and local firms based on the technological progress will provide high-technology industry to induce an attractive technological base. In doing so, the technopolis plan would be a new measure for regional development, reconstruction or redevelopment in many areas.

Background of the Technopolis Plan

Japan can be regarded as a centralised country. The most important industrial and cultural centres are concentrated in the three metropolitan areas of Tokyo, Osaka and Nagoya. Until the beginning of the 1970s, Japanese people moved into big cities in order to improve their chances of employment and companies preferred Tokyo, as almost all important political and economic institutions are located there. Because of this, the background of the technopolis plan will be examined in three different fields: technological policy, economic policy and social policy.

Background to the Technopolis Plan in the Technology Policy

Since the 1970s, advanced industrialised countries have gradually lost their grip on expansion and competition for occupation of new emerging growth fields has taken place. New technologies have played an important role in the dynamics of the restructuring process: These new technologies were involved in the rationalisation and modernisation of existing production branches and creation of new future industries which could compensate for the loss of economic dynamics in the traditional industries (Welsch 1987).

Under these domestic and international economic conditions, the Government has made attempts to relocate leading industries and to build new factories in the country since the 1960s. The government passed the New Industrial City Law (1962) and the Law for Regional Development (1964) in order to create an industrial centre for the manufacturing and petrochemical industries on the coast. In this programme the government played the main role.

For the construction of industrial centres on the coast, the Government invested immense capital that was used to build up sea ports, highways and other infrastructures. In the beginning of the 1970s, the former Prime Minister, Kakuei Tanaka, planned an archipelago with a network of '*Shinkansen*' (bullet trains), super-highway and telecommunications. However, this project failed on account of environmental problems, capital shortage and the two oil crises.

During the first oil crisis, the equipment industry reached a very low growth rate and instead, electric machines, industrial robots, etc. became the fast-growing industry sectors. These sectors needed micro-electronic technology. For this new technology, the Government invested enormous capital in the field of R&D, and stimulated a policy for regional development, which should concentrate on high-tech industry. For high-tech industry, Japan needed two different forms of R&D: first, in the field of manufacturing industry to further develop the existing production system with high-technologies; second, knowledge-oriented R&D to assist in co-operation between the science parks, institutes in industry and joint research institutes and universities (Imai 1986).

For the knowledge-oriented R&D, MITI had announced the technopolis plan in the beginning of the 1980s. Compared to both plans in the 1960s and the 1970s, the technopolis plan stressed creative technologies, education system and new information networks (Tatsuno 1986).

Japan realised that Tsukuba, 'the city of brain', as a high-ranked science park, had a relatively inferior cultural atmosphere compared to the Stanford Industrial Park, The Research Triangle in North Carolina, Sophia Antipolis, South Ile in France and Louvain University Science City in Belgium. The Government built up the Tsukuba Science City in 1965 in order to develop new technologies. It was fully supported by the Government. Its assignments were specific researches which were charged by the Government. For this reason, secret research was awarded to the Tsukuba Science City instead of private research institutes and foreign research institutes. Such a secret research policy in the Tsukuba Science City caused sceptical opinion amongst the population who disagreed with the goal of the industrial policy for spreading new technologies into every sector of society. Many Japanese people feared that the Tsukuba Science City could follow the model of American military research. Normally, military research needs enormous capital and the result of this type of research is not particularly useful to industry or society.

In order to avoid these problems and to apply basic research direct to industrial production, the conflict of the Tsukuba Science City was balanced through the technopolis plan. In principle, fundamental differences exist between the Tsukuba Science City and the technopolises: the Tsukuba Science City is a state research institute and the Government supports it fully. In contrast to the Tsukuba Science City, the technopolis plan is a programme for regional development, which should be launched by the Government through tax incentives, advice, etc. According to the technopolis plan, private industries are to take an active part in R&D programmes, in order to improve co-operation between private industries, state research institutes and universities. Compared to the technopolises, private industries play a limited role in the Tsukuba Science City. The Tsukuba Science City concentrates on basic research while the technopolises work on commissioned research, which especially aims at the commercialisation of new technologies and enables high-technology transfers from basic research (Tatsuno 1986).

Japan will manufacture technology-intensive and high-value-added products to avoid international trade conflicts with its major trade partners because these technology-intensive products account for a relatively small part of the export market and bring great advantages in the technological competitiveness around the world (Japan External Trade Organization 1983). In addition, the Government recognises that high-tech centres in the peripheral regions lead to the re-vitalisation of surrounding areas, not only through the generation of new employment but also through the diffusion of new technology into existing local industry (Witherrick and Carr 1993).

Background of the Technopolis Plan in the Economic Policy

As mentioned earlier, the Japanese economy was on the way to industrial restructuring at the end of the 1970s. Primarily, both oil crises had caused a shift in the economic system from the capital-intensive heavy and chemical industries to the knowledge-intensive industries such as micro-electronics, computers, etc, in order to lower energy costs and to improve the environmental situation. Second, Japan needed to improve its open domestic market due to the trade imbalance with its major trade partners and pressure from them.

Under these domestic and international economic conditions, the Government and companies launched their new economic policy in the 1980s. It aimed at a new industrial revolution in the technology of manufacturing, the concentration of the service sectors and the expansion of activities for research and development. In order to reach these goals, the government started to adopt the technopolis plan, which could contribute to regional development on the one hand and to strengthening the competitiveness of domestic products in the regions on the other.

Economic Background in the View of the Government

As Japan had economic difficulties on account of both oil crises during the 1970s, MITI was interested in 'Silicon Valley' in the United States. Since industrial policy did not function well by the end of the 1970s, MITI perceived its insufficiency. During this time, industry faced massive criticism by the Japanese people due to environmental problems and sun-set industries. Especially, medium- and small-sized companies experienced economic difficulties under these circumstances as the cost of construction rose constantly. They provided about 90% of the total employment in Japan.

For the sake of this bad economic situation, MITI and the private enterprises agreed that Japan should strive for a new industrial policy. It required highly developed technologies and new places of employment. Study trips to Silicon Valley (Klondike approach), North Carolina (scattered development approach) and Oregon (lab-in-the-forest approach) were arranged and MITI attempted to select one model as a Japanese strategy (Steffensen 1995). In spite of the agreement between MITI and the private enterprises, the Japanese people were very sceptical about a study by the Hamamatsu Research Team, which recommended Silicon Valley as a model for highly developed information technologies. The cautious behaviour of MITI was based mainly on unsuccessful results from the Tsukuba Science City, which had required an investment of approximately 5.5 billion dollars. Despite such a vast investment, it could not stimulate enough industrial development. Because of this, MITI was too cautious to carry out the new industrial policy, although it knew the necessity of industrial restructuring in Japan.

MITI created a state research group in 1979 to explore the possibility of installing a Silicon Valley in Japan. The president of the research group, Professor Takemochi Ishii at the University of Tokyo, attempted to build a Japanese Silicon Valley where researchers from industry and the Government could exchange their opinions and ideas regularly. With regard to the Japanese Silicon Valley, it is considered that the two highly developed technology centres, Tsukuba Science City and Tokyo, hindered the development of industrial creativity on account of exceedingly strong competition and the secret research policy.

In order to abolish these problems, the mayor of Kurume on the southern island of Kyushu, Toshiyuki Chikami, suggested the Government cause a rebirth of regional cities as technology centres. According to his idea, medium-sized cities demanded more support from central government and, in doing so, the technological gap between big cities and regions could be balanced. This idea was realised as the technopolis plan (Tatsuno 1986).

Economic Background from the View of Industry

> Japanese political centre suffers constantly from the shortage of land as a result of high population concentration already from the beginning of the 17th century, during the Tokugawa regime. At that time, a project for the land reclamation on the coast within Tokyo Bay... The Japanese present problem of overpopulation and land shortage has at least 400 years long historical backgrounds. (Gandow 1987, p.B4)

After the Second World War, the problem of population concentration in the metropolises of Tokyo, Osaka and Nagoya had reached a new dimension within the economic development. During this time a rural exodus occurred, because the people in the countryside could not find any work. On one hand, the rural exodus contributed to the economic development through low wages, and, on the other, it caused the problem of over-population in the big cities. For example, about one quarter of the whole population, approximately 120 million people, resides at the present time in Tokyo, which amounts to less than four per cent of the whole land surface.

By reason of the extreme concentration in the big cities, the price of land has risen in these three metropolises constantly since 1971. This tendency changed dramatically with the internationalisation and increased information exchange between Japan and the rest of the world in the 1980s. According to government statistics, the price of land rose 68.6% in Tokyo and about 46.6% in Osaka and Nagoya between 1987 and 1988 (Makino 1989). An extreme price rise reached to 300% in the centre of Tokyo between 1983 and 1988. This rocketing land price raised the production costs of industry, and the increased production costs weakened the competitiveness of products on the world market.

Ultimately, the enterprises started to think about moving out of Tokyo. Many companies planned to leave the three metropolises and build new premises nearby. In addition, the Government presented a settlement plan in 1985, in order to support the relocation of enterprises in the different industrial sectors in to the satellite cities – such as Yokohama, Kawasaki, Tsukuba, etc. – as much as possible (Makino 1989).

Background of the Technopolis Plan in the Social Policy

The knowledge-intensive industrial structure and regional development were the most important motives for the technopolis plan. It was also planned that the concept of technopolis would contribute to hindering the movement of the population in to the big cities and that high-technology industry could fit in with the increasing internationalised economy (Takazawa 1984). In fact, non-native residents of the large metropolitan areas wished to return to their home prefectures in order to pursue a more comfortable life than in the big city areas, if their ambitions could be fulfilled in their home towns (Japan Industrial Location Center 1990b).

Properly speaking, the idea of technopolises has been developed to balance economic and technological inequalities between the big cities and the rural regions. While the idea of technopolises sprouted throughout the entire country, local governments were willing to participate in the technopolis plan in order to be able to revitalise their regions and to create new places of employment. MITI also attempted to strengthen the technopolis plan as a policy of decentralisation. Additionally, it calculated that the new decentralising policy could minimise the economic and technological inequalities between the metropolises and the regions. In doing so, industrial capacity could be improved generally (Imai 1986).

Goals of the Technopolis Plan

The technopolis plan emerged at the time of conflicts with Japan's major trading partners over its huge trade surpluses. In the circumstances the Government attempted to find a new direction in order to avoid such trade frictions and to strengthen its competitiveness in the field of high-technologies on the world market.

Japan hopes that the technopolis plan will help in generating new creative products and thereby revise its bad reputation as a technology-imitating country (Japan External Trade Organization 1983). Simultaneously, the government aims at being an equal in basic research and industrial technologies in comparison with other advanced industrialised countries. In addition, it expects that the development of such technologies will further improve the living standards in rural areas (Economic Planning Agency 1984).

Goals of the Technopolis Plan in Different Areas

The technopolis plan has targets in three different areas: in the area of industry, to further develop the level of industrial technologies, especially high-technology industries such as micro-electronic, computer technology, biotechnology, energy technology and new materials; in the area of R&D, to solve the problem of technology transfers and produce new high-technologies; in the area of the environment, to create better living conditions with high-technology and hold the balance between the regional culture and technology (Takazawa 1984; Uchiyama 1986; Steffensen 1995).

In order to reach the goals of the technopolis plan, Japan should co-ordinate the areas of industry, R&D and environment. The technopolises stimulate the regional industries positively and enable development of products which are based on high-knowledge-intensive and information-intensive values (Takemori 1988). Furthermore, MITI recommends that private enterprises, universities and research institutes participate in the technopolis plan, and thereby the high-technology can be transferred easily to the medium- and small-sized companies (Imai 1986).

Effects of the Technopolis Plan in the Regions

The strategy of the technopolises is to be a means for the balance between the high-technology industry, R&D and the environment. The technopolis plan enables the development of industrial technologies in the regions. It influences not only regional industry, but also universities, institutes and the environment in the regions (Takemori 1988). In fact, interchange has been actively promoted between industries, universities and governmental organisations to improve the technological standard of the local enterprises. This has been carried out in various forms, such as social gatherings, study meetings, technology interchange plazas and joint researches. Additionally, the foundation of research laboratories and the expansion or enhancement of public testing and research institutions have contributed to strengthening the regional R&D and information functions (Japan Industrial Location Center 1990b). For example, the technopolis Oita, which is established on the southern main island of Kyushu in a 1,230 square kilometre area, has three principles: decentralisation, which can harmonise with the regional environment and tradition through the technopolis plan; co-existence with agriculture, forest and fisheries; development of human resources (Look Japan, Nov. 1984). Governor Toru Hirai in Yamaguchi in the southern part of the main island, Honshu, also accepts decentralisation because the technopolis plan makes it possible for local governments to improve their autonomy. Coincidently, he believes that central government should build up more universities and research institutes in the regions.

The Japanese Parliament passed the Technopolis Law in April 1983. It was aimed at a new distribution of the population and public welfare in the poorer regions. As implied above, the technopolis plan is a programme for national welfare as well as for the development of high-technologies (Tatsuno 1986).

In relation to the technopolis plan, Morihiro Hosokawa, the former Governor of Kumamoto in the southern island of Kyushu, set the long-term goal as follows: in the 1990s Japan must pursue creative research in information systems, mechatronics, biotechnology and new materials; Japan will become a technological leader in the world; the technopolis plan will create internationally competitive industries, creative technologies and an attractive urban environment.

The Paradigm of Technopolis

In this chapter, the paradigms of the technopolis are presented in order to explore how the technopolis plan is developing in the regions and which problems are hindering it. The model technopolises are chosen randomly and I have attempted to select them with a regional division.

Technopolis Hakodate

The northern main island, Hokkaido, has about 22% of the entire Japanese area. However, the population of Hokkaido is only 5% of the whole Japanese population. The colonisation of Hokkaido began relatively late. With the suppression from the West, especially on the part of the United States of America, the sea port of Hakodate was opened in 1859. Since then it has concentrated on ship-building and pharmaceutical industries.

The sea port Hakodate has played a part in the connection between the main island, Honshu, and the island Hokkaido because it is located between the straits 'Tsugaru'. For these geo-political and geo-economic reasons the central government supported the island Hokkaido strongly. With strong financial support from central government, the resettlement of its population on the island amounted to five million (Japan External Trade Organization 1983). With this historical background, Hokkaido is regarded as the last frontier and the technopolis plan in Hakodate is also focused on strategical significance.

The technopolis area is about 962 square kilometres and in this area four cities, Hakodate, Kamiso, Ono and Nanae are located. Hakodate is the mother-city of the technopolis. Hakodate already has well developed ship-building and fishing industries and aims at increasing its industries, building industrial complexes for high-technology and creating an attractive environment in the city.

The technopolis Hakodate attempts to carry out its assignment with the motto 'Improvement of own industry and application of local resources in the region'. The governor of Hokkaido, Takahiho Yokomichi, stresses the priority of the technopolis plan in Hakodate as follows: 'it is important to refine the technological capabilities of existing industries in addition to inviting high-tech industries to locate facilities in the technopolis.' (Look Japan, June 1985)

In order to reach the goal, three organisations were built up in 1984: 'The Committee for Promotion of Technology in Hakodate' (Hakodate-ken Gijutsu

Shinko Iinkai), which encourages technical cooperation between different industries and universities; 'The Conference for Industrial Invitation in Hakodate' (Hakodate-ken Kigyo Yuchi Suishin Kyogikai), which arranges the resettlement of industries into the technopolis Hakodate; 'The Association for Promotion of the Technopolis Hakodate' (Tekunoporisu Hakodate Gijutsu Shinko Kyokai), which consists of representatives from the local government and finances and advises on R&D projects and the building of new industrial facilities which are suitable to the region. The Centre for Regional Industry Technology was opened in 1986. This centre contributes to the application of high-technology to the regional industries. For example, it introduced the high-technology, 'mechatronics', which is installed in the ship-building and fishing industries for the improvement of existing regional industries. Additionally, biotechnological research is under way in order to utilise the sea resources more effectively (Japan Industrial Location Center 1990b). It should also be added that special incentive measures regarding taxation and financing are available for companies settling in the technopolis area (Committee for the Promotion of Investment in Hokkaido 1992).

For the development of existing industries in the technopolis Hakodate three different industrial complexes were constructed:

(1) The industrial complex for marine (Kaiyo Kanren Sangyo-gun) is the centre of high-technology for the development of the marine industry. The North Marine Research Centre is responsible for the R&D project.

(2) The industrial complex for social development (Shakai Kaihatsu Sangyo-gun) is engaged in the development of telecommunication and medical equipment. The centre for urban development contributes to carrying out R&D programmes in the complex.

(3) The industrial complex for application to resources (Shigen Katsuyo Sangyo-gun), concentrates on the development of agriculture, forestry and mining. The centre for application to resources carries out R&D programmes in the complex.

With the enlarging of industrial complexes in the technopolis area, the industrial shipments are projected to increase from 287.2 billion Yen in 1980 to 450 billion Yen in 1990. Additionally, the number of employees is targeted to increase from 17,417 to 19,000 in the same period (Look Japan, June 1985). (See Table 8.1) These targets are partially reached; the target of industrial shipment in 1989 was even exceeded. However, the target of new employment seems to be unsuccessful (see Table 8.2)

The role of the University of Hokkaido, especially the department of fishing sciences, is also very important in the technopolis Hakodate. This department contributes to improving technological standards in the technopolis and co-operates with regional industries. Also, the industrial infrastructure in the technopolis was essentially improved through the construction of the Seikan-Tunnel in 1985, which connects Hokkaido with the main island of Honshu. The technopolis

Hakodate is preparing its own future and is ready to play the appropriate role as the last frontier in Japan.

Table 8.1 The development of the Technopolis Hakodate

	1984	*1985*	*1986*	*1987*	*1988*	*1989*
Industrial Shipment (mil. Yen)	313,568	308,419	308,131	308,016	321,942	341,538
New Employment	650	531	462	491	511	444
New Investment (10,000 Yen)	286,723	333,578	130,812	151,729	208,880	437,244

Source: Jun-Ichi Ochiai, Director of New Technology Industry Division, Department of Commerce, Industry, Labor and Tourism, Hokkaido Prefecture, 1992

Table 8.2 The targets of Technopolis Hakodate (at 1984 prices)

	1984	*1989*	*1995*	*2000*
Industrial Shipment (mil. Yen)	287,200	339,000	524,700	729,700
Employment (Person)	17,417	16,239	19,530	21,700
Population (Person)	380,517	381,642	420,000	440,000

Source: The Association of Promotion for the Technopolis Hakodate. 1993

Technopolis Ube (Yamaguchi)

The region of Yamaguchi, on the western tip of the main island of Honshu, has a powerful political heritage in Japanese history. Yamaguchi challenged the central government and, under the Ouchi Family (1336–1550), it was the political and cultural centre on the west side of Japan. After the Mori Family (1550–1868) took over, Yamaguchi continued to maintain its political power as the centre of the west of Japan. Later, it was defeated by the central power of the Tokugawa Shogunat in 1868. From this tradition of political struggle for central power, it was possible for Yamaguchi to produce Japan's first Prime Minister (Hirobuumi Ito) and the two Prime Ministers, Nobusuke Kishi (1957–1960) and Eisaku Sato (1964–1972), in the post-war period (Tatsuno 1986). These political conditions played a very important role in Japan and brought advantages for building up the technopolis Yamaguchi.

The main industry of Yamaguchi until the beginning of the 1950s was mining. With the application of oil as the main resource for industry, Yamaguchi produced petroleum, cement, steel, glass, etc. These products account for 65% of the entire industrial output in the region and this share is in the top position for the whole of Japan. In spite of such a capacity, productivity and employment dropped continuously due to the need of vast capital and energy resources by the heavy industry. In order to solve these problems and to bring new high-technologies to the region, Yamaguchi struggles for the technopolis plan (Japan External Trade Organization 1983).

The technopolis Ube (Yamaguchi) is composed of eight cities: Ube, Yamaguchi, Onoda, Mine, Ogori, Ajisu, Kusunoki and Sanyo. Katsuji Yuda, Director of the Technopolis Plan, explains the reason for the wide technopolis area and the historical aspects: 'unlike other prefectures, Yamaguchi does not have one major city, but many small cities. This pattern was caused by the political fortunes of the Mori Family.' (Tatsuno 1986) From this background the structure of decentralised population arose in the region and it resulted in the relatively reasonable price of land which enabled the multi-centralised technopolis to be built. In the beginning of the 1980s, the land price was 28 dollars per square metre in the technopolis area. According to the statistics of the Yamaguchi region, it was 30% lower than in the rest of west Japan (Japan External Trade Organization 1983).

To be attractive to high-technology-oriented companies outside the region, Yamaguchi improved its infrastructure with the techno-road plan, which connects the existing super-highway, the bullet train (Shinkansen), the Yamaguchi-Ube airport and two sea ports. Thereby, the technopolis Ube (Yamaguchi) can reach a network in the technopolis area. It expects an increase of population in the area from 408,000 in 1980 to 470,000 in 2000 and set its research priority in the field of microelectronics, new materials, mechatronic, biotechnology and energy technology (Ube Technopolis Construction Promotion Council 1993).

This technopolis plan will require immense capital – similar to other technopolises. It will cost approximately 1.3 billion dollars. In order to minimise the financial burden in the region, Yamaguchi is embarking on a long-term technopolis plan. Its realisation will take 18 years which is 10 years longer than MITI planned. According to MITI's plan, the first step of the technopolis plan, the construction of basic infrastructure, would be finished in 1990 at the latest. It belongs to one of 20 technopolises which have completed their first step. The second part of Ube Technopolis Development Plan was approved in April 1992 and the Technopolis Ube began constructing Ube New Town in October 1992 (Park 1994).

In relation to the financial plan, the technopolis planners explain that 24% of the entire cost is financed by the Yamaguchi region. The small cities contribute 20% and the local industry 21% of the cost. The rest will be subsidised by the central government and state-owned corporations (Tatsuno 1986). Compared with other unfavourable technopolises, the technopolis Ube (Yamaguchi) expects more subsidy from central government because it has better political connections.

The big disadvantage of the technopolis Ube (Yamaguchi) is a lack of high-qualified researchers and scientists. Usually they are engaged in research in the metropolitan areas, such as Tokyo, Osaka and Nagoya. For this reason, the technopolis Ube (Yamaguchi) set up the Techno VIP system as an invitation programme. It invited highly qualified researchers and scientists to the technopolis area and the Techno-VIP system is carried out by the organisation of development for industrial technology built in 1983. In addition, this organisation takes care of the improvement of technological standards in medium- and small-sized industries. It offers companies not only a reasonable land price but also well-arranged living areas. To take this opportunity, the vehicle manufacturer Mazda has opened a factory and NEC invested 150 million dollars to produce IC Chips. The electronic company NEC created 600 new jobs in the region and the number in employment rose to 900 within three years. Other new companies, such as Japan Kanizen and Yamaguchi Computing Center, have also resettled in the technopolis Ube (Yamaguchi).

Through the construction of technopolis Yamaguchi, expectations are of an increase in industrial output from about 1,115 billion Yen in 1980 to about 2,410 billion Yen in 1990 and approximately 14,000 new jobs by 1990 (Look Japan, Dec.1984). (See Table 8.3) However, these expectations seem to be optimistic and extremely over-estimated if we look at the statistical background. During the period between 1980 and 1987, the industrial products and the number in employment in the technopolis area decreased 5.9% and 4.6% respectively. Thus, its aggregate order of precedence was 16th out of 20 technopolises in the same period (Park 1994).

Table 8.3 Objectives in the second term of the plan

	1980	*1985*	*1995 (Projected)*	*2000 (Projected)*
Industrial Shipment (100 mil. Yen)	11,149	10,408	17,450	21,340
Employment (Person)	31,311	30,541	31,980	32,470
Value Added Product (mil.Y/Person)	9.9	9.3	16.3	22.4
Population (1000 person)	409	426	453	467

Source: Ube Technopolis Construction Promotion Council, Yamaguchi City, 1992

Technopolis Oita (Kenhoku Kunisaki)

The technopolis Oita (Kenhoku Kunisaki) is located on the west side of Oita airport. The technopolis area is the centre of Japanese Buddhism, which began originally in the era of Nara and Kyoto (794–1185). Therefore, there are many buddhist temples and statues. A few statues are carved on large stones. To protect these cultural icons, the technopolis Oita (Kenhoku Kunisaki) is built up in five different areas, such as Nakatsu, Usa, Takeda, Kunisaki, Kitsuki.

Two mother cities, Oita and Beppu, are located about 50 kilometres from Oita airport and in both cities heavy industry, such as oil refinery, petrochemical industry, steel industry, etc. were equipped during the 1960s and the beginning of the 1970s. Since then the technopolis Oita has concentrated on the new high-technology, such as microelectronics, biotechnology, etc., instead of heavy industry because the product of new high-technology is high-value added. It also fits well with air transport, which can deliver products to the market faster than any other transport method and thereby contributes to maximising the profit of the firms (Japan External Trade Organization 1983).

Originally, the technopolis plan was one of five regional development plans in Oita connected with high-technology and the regional economy. The other four plans are: the Oita District New Industrial City Construction Project; Marinepolis Project; the Agriculture and Tourism Development Project and Hita Kusu Model Settlement Zone Project.

As Oita introduced the new industry project in 1979, which aimed at the high value added products and pre-supposed an airport located a maximum of 50 kilometres away from the industry parks, MITI offered Oita the opportunity to be a model technopolis. Oita had regional initiatives before the central government announced the technopolis plan for regional development. This future-oriented industrial policy in Oita resulted from the prominent performance of the governor, Morihiko Hiramatsu, since 1979. He worked as Director of the Department of Electronics and was the main figure on the committee of technopolis 90 in MITI before he was elected in his home town of Oita Prefecture (Castells and Hall 1994).

Since Morihiko Hiramatsu took power in Oita, he has constantly tried to transform its industrial structure from the conventional heavy industry to the new high-technology industry on the island of Kyushu. For example, he influenced the resettlement of the American company Materials Research Corporation (MRC) in the technopolis area. MRC opened a semi-conductor factory, together with the Japanese company Modoriya Electric, in Oita and both companies obtained a 1.5 million dollar loan at low interest rates from the Japanese Development Bank (Tatsuno 1986). Furthermore, the governor invited more private companies, such as Sony, Canon, NEC, Toshiba, Dai-Nippon Ink and Chemicals, HOKS-Electronics and Kyushu Matsushita Electric, to the technopolis area.

As the technopolis planners in Oita chose a wide area for the technopolis, MITI protested heavily against the plan because it suggested a new industry park near to Oita airport. After long discussions between the local government and MITI, the technopolis planners in Oita continued with their own decentralized plan. According to the opinion of the technopolis planner, Suenobu Tamada, MITI could not understand the regional situation relative to the special requirements of the Oita region or the wishes of the inhabitants. The designed technopolis area is composed largely of farm land and the buddhist cultural centre. For this reason, the primary goal of the technopolis Oita is harmony between the new industry and the rural atmosphere.

The technopolis planners in Oita preferred the resettlement of small- and medium-sized factories in the technopolis area because they could use the existing infrastructures without any additional financial burden on the region and minimise the environmental pollution. The best location for high-technology-oriented small- and medium-sized firms is where the worker could arrive within 30 minutes by foot. For example, the American company, Texas Instrument, produces micro-chips in the technopolis area which are produced mainly by Japanese house-wives from surrounding towns. They have reached the highest quality among 50 other factories of Texas Instrument stationed abroad (Japan External Trade Organization 1983).

The firms located in small towns also profit by the advantage of the low land price and low labour cost, and the high-technology-oriented companies in the small towns reduce the rural exodus into the metropolitan areas. In order to further develop these advantages for both companies and regions the governor stressed the motto 'One region and One product', which encouraged the improvement of technological standards in the region. The products of the region should be competitive at least on the domestic or international market. According to his opinion, Oita must work regionally with the global perspective and this attempt is an internal element of balanced economic development in Oita.

Oita built a committee for the motto 'One region and One product' consisting of representatives from agriculture, fishing, finances and scholars, etc. and it considers not only unique but also potential export products and technologies. To support this committee, Oita held the first fair for high-technologies in October 1984.

The region Oita is interested in the resettlement of high-technology-oriented companies in the technopolis area and in the creation of new places of employment, as well as the direct application of high-technology to the region and technology transfer to local firms. For example, the local company Denken Engineering can produce small personal computers for schools through technology transfer. The micro-computer producer HOKS-Electronics uses micro-chips manufactured in the region for its micro-computer, so productivity has been increasing dramatically since the establishment of the company in 1982. The president of HOKS-Electronics, Michinori Kudoh, says

> We are a locally-capitalized venture business which brings brains and technologies back to Oita from Tokyo and foreign countries, and we would like to lead other high-tech businesses in the construction of the Oita technopolis... Kyushu has many plants producing ICs, which account for 40% of Japan's total production and 10% of the world's. We call ICs the rice of industry. But although they produce it, industries in Oita do not eat this delicious rice. They ship most of it to Tokyo, Osaka and foreign countries. But HOKS Electronics acts as a user of these IC's. (Look Japan, Nov.1984)

With the increasing resettlement of high-technology-oriented firms in the technopolis area, the productivity of existing small- and medium-sized local companies has been improved. In the technopolis area industrial output was raised from about 201 billion Yen in 1980 to about 647 billion Yen in 1990 (Look Japan, Nov.1984). (See Table 8.4)

The completion of the required infrastructure, facilities for R&D and living areas, etc. for the technopolis Oita was reached, as planned, in 1990 and the technopolis Oita further developed its goals of decentralisation, co-existence between the existing agriculture and the high-technology-oriented industries and the development of human resources.

Table 8.4 Technopolis progress report in Oita prefecture

	Base Year 1980	*1988*	*1989*	*1990*	*1991*	*Projected 1995*
Industrial Shipment (mil. Yen)	198,144	479,868	535,327	608,858	707,171	885,300
Number of Employees in Manufacturing	17,455	21,634	22,546	23,733	26,307	26,966
Population in Technopolis area	281,513	279,289	277,601	276,759	275,569	278,000

Source: Mitsuhiro Kawano, Director of Planning & Coordination Division Oita Prefecture, 1994

Analysis of Problems in the Technopolis Plan

Given the judgement of MITI in November 1992, the result of the first term (1983–1990) in the technopolis plan – measured by the shipment of manufacturing products, the manufacturing value-added, the number of manufacturing workers and the population – is considered as successful. The development of technopolis areas surpassed the average of national development (Tsukahara 1994). (See Table 8.5)

Table 8.5 The rate of development in the target of technopolis plan (per cent)

	Shipment of Manufacturing Products	*The Manufacturing Value-added*	*Number of Manufacturing Worker*	*Population*
Nationwide	57.0	41.4	8.5	5.6
20 Technopolises	59.8	40.6	7.0	5.9
26 Technopolises	65.2	45.3	10.3	6.0

Source: A Process of Technopolis Construction, MITI November 1992

Note: $\text{The Development Rate} = \frac{\text{Amount of Product in 1990} - \text{Amount of Product in 1980}}{\text{Amount of Product in 1980}}$

However, it varies heavily at local levels as Table 8.6 shows. The reasons are over-expectation on the technopolis plan since its start, a drastic change in the economic situation due to the rapid appreciation of the yen against the US dollar in the second half of the 1980s and the problems for local governments in carrying out their own policies because of insufficient co-ordinated planning (Hiroshi 1994). Consequently, the attempts of ambitious local governments to attract high-tech-oriented companies failed, especially in areas distanced from the three metropolitan areas. The case of Kyushu Island is exceptional since it provides ample low-cost female workforces and the clean air and water which are necessary for IC-Chip industry. Hence, the technopolis concept was widely recognised as a policy tool to minimise the domestic industrial deficiency resulting from an increase of Japanese Foreign Direct Investment. Finally, the technopolis plan was formally revised and it's target year for completion was prolonged from 1990 to year 2000. In addition, it is regarded as the super-pioneer of the multi-polar patterned land use policy based on a long-term public development project (Steffensen 1995).

Given the outlines of an advisory council report issued by the Japan Industrial Location Center in 1990, several problems still remain on account of general socio-economic structures such as Tokyo-oriented infrastructure networks, the existence of qualified work-force mainly in the three metropolitan areas, etc.

Table 8.6 The achievement ratio of the technopolis plan in the first term (each technopolis)

	Over 100%	*80%–100%*	*50%–80%*	*0%–50%*	*Less than 0%*
The Shipment of Manufacturing Products	3	3	5	8	0
The Manufacturing Value Added	0	1	3	14	1
The Number of Manufacturing Workers	1	3	3	7	5
The Manufacturing Value Added per Worker	1	1	4	13	0
Population	0	2	5	9	3

Source: Survey by Prefecture in 1992

Thus, in order to reach the goals of the technopolis plan, a further build-up of organisational infrastructure, adequate educational facilities and creative environments must be established in the technopolis areas (Steffensen 1995). In addition, the report provides structural conditions and planning strategies in two terms which consist of the first (1983–1990) and the second (1991–2000) (Senta 1990). (See Table 8.7).

Table 8.7 Overview of structural conditions and planning strategies

First Technopolis Period	*Second Technopolis Period*
Given Conditions:	**Given Conditions:**
Industrial structural switch from heavy and large scaled fordistic production to light and small-scaled flexible manufacturing systems	Increase of enterprises locating overseas along with the appreciation of the yen/ progress of industrial structural adjustment
An approach run towards a technology nation (gijutsu-rikkoku)	Progress of globalism
A period of financial reconstruction	A wave of technological innovation and international competition on technological development power
Demands for the revitalisation of local economies	Reinforced foundation of technological development power (maturing of the first technopolis period)
Enlarging disparities of the technological capabilities between regions	New information technologies and spread of informationalisation into local economies
Decline of local industries, concentration of high-tech industries in large city areas, enlarging regional disparities	The advance of soft and service economies
	Diversification and change of value and consciousness among the Japanese
	Recovery of national financial power
Basic strategy of first period:	**Basic strategy of second period:**
Reinforce the foundation of technological development capabilities	Lay stress on endogenous modes of development
Transfer to high-tech local industry	Give prominence to individualisation
Dig out local needs and make the most out of the local seeds	Formation of technopolis networks
Technological innovation from the grassroots	Accomodate to internationalisation (open policy, technology transfer, networks)
Respect independency of localities, attach to the soft-technology base	Implement a city making machi-zukuri consisting of san, gaku, ju, and yu

Source: Nihon Ritchi Senta, 1990, p.50

Conclusions

After World War II, Japan was in a chaotic economic situation: high inflation rate; low productivity of industries; high unemployment rate and so on. It resulted in the Government creating industry and technology policies. MITI planned and carried out these policies, which concentrated on the reconstruction of basic industries such as steel, ship-building, coal energy and fertilizer in the 1950s and focused on rapid economic growth through strategic industries such as vehicle and electronic industries in the 1960s. Following the first oil crisis (1972–1973), Japanese industry was shifted from capital-intensive to knowledge-intensive through the market mechanism and limited government guidance. Besides, Japan has continuously developed a corporative system, in which the Government, industry and the scholar could arrive at a consensus on the industrial policy. The techno-corporative system is regarded as a result of government intervention policy and even criticised as a product of Japan by the West. However,Japan argues that it is a part of Japanese culture and that government intervention in the economy, especially in the private sector, has been very limited since the 1970s. Nevertheless, the role of government is still important, to guide the future direction of industry. It must be noted that the success of the economy results from aggressive business management operating in a growth-oriented system and the appropriate government intervention for resource allocation to strategic industries.

With a background of Government intervention in the economy, the technopolis plan was announced by MITI in order to solve existing industrial and technological problems on one hand and to minimise the impact of the domestic market opening on the other. As MITI introduced the plan in 1980, it intended two or three technopolises. However, the number of technopolises has been raised to 26. It shows that the regions are fighting for their development.

Japan plans that the weakness of basic research, compared to Western countries, should be balanced through the technopolis plan and it keeps the technological standard in the top position for the future. For that, co-operation between the government, industry, and scholars is necessary. In spite of these efforts, the technopolis plan does not show any strong regional success because it is still in a period of development. For example, 20 of the entire 26 technopolises have reached their first step by completing the infrastructure, facilities for R&D activities and so on, in 1991. Actually it is too early to conclude the result of the technopolis plan. It will start to come out slowly at the end of the 1990s, though the central and local governments, companies and research institutes operating in the technopolis areas are convinced that the technopolis plan contributes to improving technological development and to generating new places of employment in the regions. For this reason, I believe that if it can realise regional development on time, it will be one of the most important regional policies in Japanese history.

References

Castells, M. and Hall, P. (eds) (1994) *Technopoles of the World, the Making of 21st Century Industrial Complex.* London: Routledge.

Gandow, A. (1987) Japan, Daten, Information, Analysen, Handelsblatt, 11.11.

Hahne, U. (1985) *Technologieparks.* Bonn: Deutscher Industrie and Handelstag.

Imai, K-I. (1986) *Japan's High Technology Industries.* Tokyo/Washington: University of Washington Press.

Japan External Trade Organization (1983) Now in Japan, Nr.34

Japan Industrial Location Center (1990a) *Technopolis Policy of Japan.* Tokyo: JILC.

Japan Industrial Location Center (1990b) *Present Situation of Technopolis and Techno-Network.* Tokyo: JILC.

Kodansha (1993) *Japan: An Illustrated Encyclopedia.* Tokyo: Kodansha.

Look Japan, Sep.10.84, Nov.10.84, Dec.10.84, June.10.85.

Makino, N. (1989) *Prognoses of the 1990s in Japan.* Tokyo: Mitsubishi Research Institute.

Nishioka, H. (1989) *High Technology Industry: Location, Regional Development and International Trade Friction.* Tokyo: Academic Press.

Park, S.-C. (1994) *Technopolises in Japan.* Hamburg: Dr. Kovac.

Senta, N.R. (1990) Research Study on the Technopolis Drive, Technopolis 2000 Concept Check-up-Report, March.

Steffensen, S.K. (1995) 'Techno Obscuring Polis: A Decade of Public Technopolis Planning in Japan.' *Japan Forum 1,* 67–83.

Takazawa, N. (1984) *Cities of the High-Tech Era.* Look Japan, Oct.10.

Takemori, K. (1988) Impacts of microelectronics on human resource development in production systems, an empirical research of factory automation in Japan. In T. Davis and M. Mizuno (eds) *Employment Problems Under the Conditions of Rapid Technological Change.* Berlin: Duncker and Humbolt.

Tatsuno, S. (1986) *The Technopolis Strategy. Japan, High Technology, and the Control of Twenty First Century.* New York: Prentice Hall Press.

Toda, T. (1990) The location of high technology industry and the technopolis plan in Japan. In J. F. Brotchie, P. Hall and P. W. Newton (eds) The Spatial Impact of Technological Change. London: Routledge.

Tsukahara, H. (1994) A Study of Evaluating the Technopolis Policy from Five Indices. *Annals of the Japan Association of Economic Geography 40,* 3, 56–63.

UBE Technopolis Construction Promotion Council (1993) *Ube Phoeni Technopolis.* Ube: UTCPC.

Uchiyama, Y. (1986) *Japans Stellung in der Weltwirtschaft.* Constantin von Barloewen/Kai Werhahn-Mees, Japan und der Westen, Band 2.

Vernon, R. (1966) 'International Investment and International Trade in the Product Cycle.' *The Quarterly Journal of Economics 80,* 190–207.

Welsch, J. (1987) MITI-sierung der Länderstukturpolitik? WSI-Mitteilungen, September 1987, p.552.

Witherick, M. and Carr, M. (eds) (1993) *The Changing Face of Japan: A Geographical Perspective.* London: Hodder and Stoughton.

CHAPTER 9

New Industrial Spaces and National Technology Policies

The Case of Kyushu and the Japanese 'Technopolis Strategy'

Rolf Sternberg

Introduction

There are many indications that, in terms of political economy, the region is experiencing a revival and – implicitly in this context – topics of economic geography are gaining importance. Michael Porter's bestseller (1990) *The Competitive Advantage of Nations* supports this assumption. According to him, national competitive advantages of individual industries are frequently owing to the occurrence of these industries in geographical clusters in just a few regions. Annalee Saxenian, in her recent book, entitled *Regional Advantage*, reaches the same conclusion. However, contrary to Porter's national economic point of view, her arguments come from the regional perspective.

In Japan, too, clusters of individual industries (for example, mechanical engineering in the south of the Tokyo-Yokohama megalopolis, cf. Takeuchi 1994) resembling industrial districts occur. However, it is not known whether technology districts or technopoles with especially close intra-regional linkages between fast-growing R&D intensive industries can also develop in spatially and economically peripheral regions of Japan. Also, Western industrialised countries have no proof that the MITI's technology policy – besides its recognised national economic and sectoral impact – had, and has, regional economic effects. For the purposes of this study, Kyushu was chosen as an example. This part of Japan, in the 1980s, seems to have succeeded in advancing from an economically and spatially peripheral to a high-tech region and created quite a sensation as 'Silicon Island'. Kyushu is, in effect, a 'new industrial place' – to use Scott's (1988) term; an industrial region which owes its emergence to technology-intensive industries, the majority of which did not settle in the old industrial northern part of the island. By Japanese standards, however, Kyushu continues to be underindustrialised. Therefore, it will also be discussed in this chapter whether or not Kyushu can be considered a technopole. Part Two of this contribution, with the help of as yet unpublished data, tries to evaluate the success of the Technopolis Programme using the six technopolis zones on Kyushu as case studies.

Regional-Economic Structure of Kyushu

Kyushu is the southernmost of the four main islands of the Japanese archipelago and is often termed '10% economy'. In 1992, the 13.4 million inhabitants (10.7% of total population in 1994) produced approximately 8.6% of national GDP. The island – comparable in size to Switzerland – shows high percentages of value-added in the agricultural and fishing industry sector, whereas the manufacturing sector (especially mechanical engineering) is underdeveloped; in 1993, Kyushu's share of Japan's industrial turnover was only 6.2% (Kyushu Industrial Advancement Center 1995; Matsubara 1992; Yada 1987). Seven of the 47 Japanese prefectures are located on Kyushu (Kumamoto, Saga, Fukuoka, Oita, Miyazaki, Kagoshima and Nagasaki); for regional economic analyses they constitute an appropriate administrative level. (Figure 9.1).

Not only the value-added shares but also other economic indicators, such as per capita GDP, productivity in manufacturing and the relationship between demand for and availability of labour, reveal that Kyushu still is a peripheral part of the country in terms of location as well as economy and considerably lags behind other large industrial regions. This general judgment, though, disregards both the older and the more recent economic history of the island and its regional disparities. Until World War II, the northern part of Kyushu, particularly around the metropolis of Kitakyushu, was the third largest industrial region with a share of 10% of national industrial production (Nishioka 1991). In the wake of the discovery of large coal deposits, numerous large establishments of the iron and steel industry settled there, later followed by chemical industries and shipyards in other parts of the island. Since World War II, Kyushu's share of employment and production in these industries has steadily decreased – especially the basic industries, which depend on cheap coal and predominated until after the War but started losing in importance when coal was substituted by oil in the 1960s and again in the 1970s due to the oil crises. For a long time, the industrialisation of the island was limited to the very north and small coastal areas in Nagasaki and Oita, the larger part by far remaining characterised by agriculture and forestry. Correspondingly, the old industrial regions in Fukuoka and Kitakyushu (iron and steel industry and coal mining) had to struggle with the problem of restructuring, which is typical of this type of region (Sternberg 1995a; Tamura 1992).

Its reputation of a 'Silicon Island' – a term first used in an article in *Newsweek* in 1984 – Kyushu acquired toward the end of the 1970s when a good many

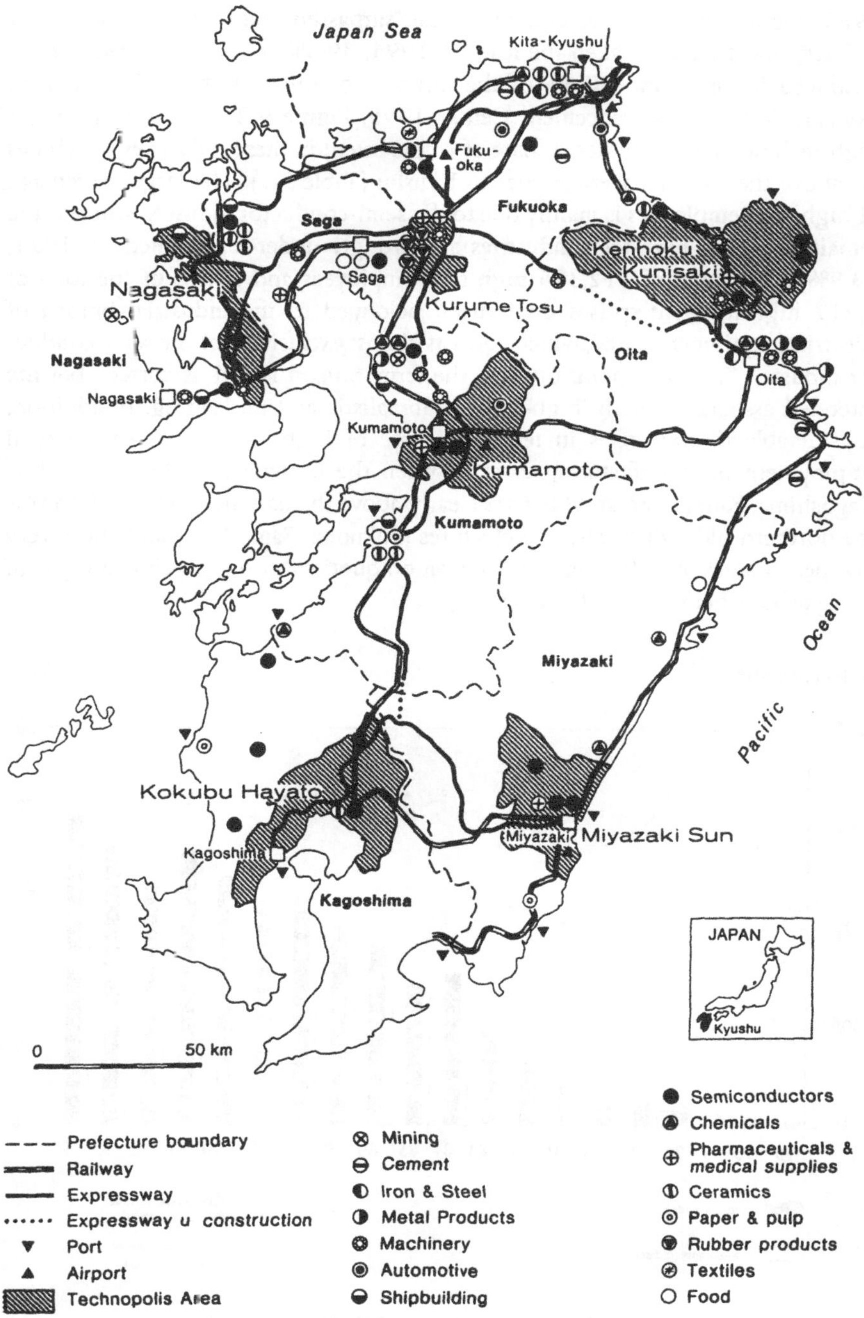

Source: Kyushu Industrial Advancement Centre 1994

Figure 9.1 Kyushu: economic structure, transport infrastructure, and Technopolis zones

semi-conductor businesses began to establish themselves in all prefectures. In 1985, the turnover of these enterprises had surpassed that of the iron and steel industry and they had taken the lead. In 1994, 39.3% of the integrated circuits produced in Japan and 30.6% of the turnover of this industry fell to Kyushu (Kyushu Industrial Advancement Center 1995; Figure 9.2 below). In all eight high-tech branches together, a share of employment in these industries which lies far above the national average goes to Kyushu (Table 9.1). The large percentage of high-tech employees is mainly due to the semi-conductor industry, whereas the remaining seven high-tech industries are strongly under-represented. In 1991, 73.9% of the total of 112,395 high-tech employees and 51.0% of the total of 1,217 high-tech enterprises on Kyushu belonged to the industrial branch of 'electrical equipment parts', which consists almost exclusively of the semi-conductor industry. From this point of view, the term 'Silicon Island' is correct, but the intended association with 'high-tech' is unrealistic and misleading. In addition, considerable discrepancies in terms of share of high-tech employees in total employment in manufacturing exist between the individual prefectures. While Kagoshima, Kumamoto and Oita are clearly above the national average of 13.0%, the northern old industrialised prefectures (Fukuoka, Saga, Nagasaki) show very low percentages. At 0.91, the overall location quotient of high-tech employment on Kyushu is below the national average.

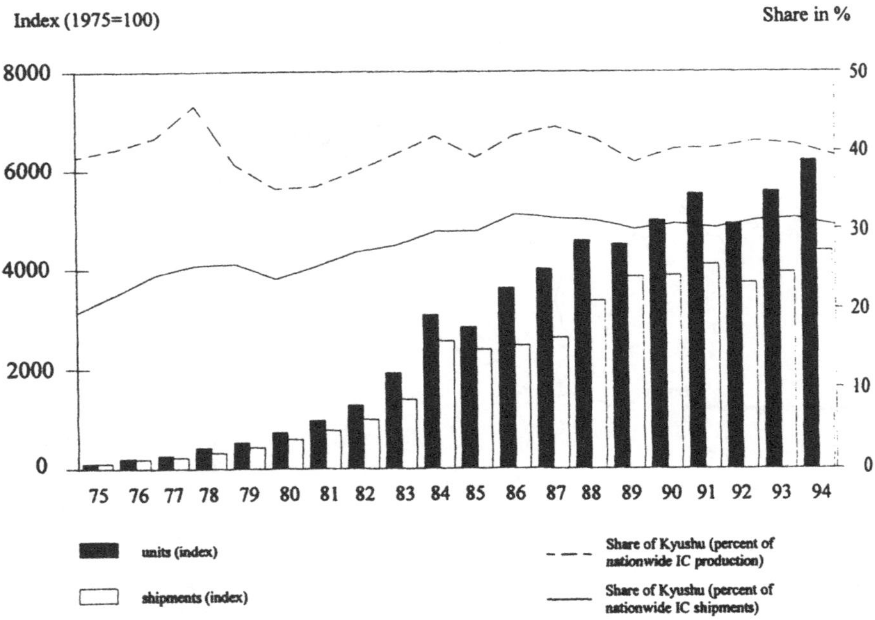

Source: Kyushu Industrial Advancement Centre 1995, Matsubara 1994

Figure 9.2 Development of IC production on Kyushu, 1975–1994

Table 9.1 Employment in high-tech industries in Kyushu prefectures, 1991

Prefecture	*Employment LQ**	*High tech industries*		
		Electrical equipment part (incl.integrated circuits)	*Other high tech industries***	*Total*
Fukuoka	Employment	18.649	8.703	27.352
	LQ	0.96	0.34	0.60
Kagoshima	Employment	19.502	894	20.396
	LQ	3.01	0.10	1.35
Kumamoto	Employment	18.124	5.383	23.507
	LQ	2.42	0.54	1.34
Miyazaki	Employment	9.231	2.175	11.406
	LQ	1.89	0.33	1.00
Nagasaki	Employment	3.959	2.563	6.522
	LQ	0.77	0.37	0.54
Oita	Employment	9.568	5.989	15.557
	LQ	1.88	0.88	1.31
Saga	Employment	4.029	3.626	7.655
	LQ	0.89	0.60	0.72
Kyushu totals	Employment	83.062	29.333	112.395
	LQ	1.57	0.41	0.91
Japan	Employment	785.652	1.051.297	1.836.949
	LQ	1.00	1.00	1.00

* Location quotient, Japan = 1.00

** including the following industries (MITI 1986); 206 Pharmaceutical Products; 304 Communication Equipment; 305 Computers and Peripherals; 306 Electronic Equipment; 307 Electrical Measurement Machines; 308 Electrical Equipment and Parts; 323 Medical Instruments and 325 Optical Instruments

Data source: MITI 1992 (Establishment Census 1991)

Between 1981 and 1991, the high-tech sector on Kyushu was very dynamic. In terms of numbers of businesses and their employment, the prefectures of Kyushu, without exception, showed considerably higher growth rates than the rest of Japan (MITI 1992). However, the population development continues to take a less favourable course on Kyushu than in the rest of the country on average. Due to unsatisfactory or lack of jobs, there is an out-migration – especially of the younger population, which even the decentralisation of semi-conductor industries has been unable to bring to a halt. From this point of view, Kyushu, even economically, continues to be a peripheral region.

Kyushu's semi-conductor industry has undergone constant restructuring. As a reaction to the revaluation of the yen in 1985, and increasing trade restrictions imposed by the Western industrialised countries, many regional semi-conductor manufacturers re-oriented their production toward value-added-intensive products (custom-made semi-conductors and mass production of large-scale integrated circuits with one megabyte storage capacity) (Sargent 1987; Matsubara 1992). A further indication of the ageing of this high-tech branch is increased automation (that is, substitution of labour by capital) which will change the requirements on quality and quantity in a way that is very likely to render Kyushu's specific locational advantages less important. While the quantity and quality of semi-conductors made on Kyushu was still on the rise in the early 1990s, a further increase in employment seems unlikely.

In the coming decade, the automotive industry (particularly the production of automobiles) is expected to take the lead, leaving the semi-conductor industry behind. The automotive industry also utilises the region's comparative advantages, namely the availability of relatively cheap and flexible labour and fully developed industrial space. Therefore, primary assembly plants are expected to settle on Kyushu and function as 'buffer plants' as long as the present changeover to the new 'Flexible Manufacturing System' is not concluded (Tomozawa 1992). The announcement by Nissan (new plant in Kanda/Fukuoka), Honda (Kumamoto) and Toyota (Miyata/Fukuoka) of the establishment of new production facilities on Kyushu did not exactly cause a run on Kyushu by subcontracting industries which need to be close to their customers. In mid-1993, there were 206 enterprises on Kyushu directly or indirectly (as subcontractors) belonging to the automotive industry and, in 1993, producing 4.9% of the national automotive output (only 3.4% in 1988) (Kyushu Industrial Advancement Center 1995).

Factors Promoting the Growth of Technology-Intensive Industries During the 1980s

Spatial Division of Labour in Japan

In order to understand the development of Kyushu's regional industrial structure it is necessary to define the term 'spatial division of labour'. Spatial division of labour is characteristic of all internationally competitive Japanese industries. The high-tech industries are well suited for explaining this term. Spatial division of labour in Japanese industry is based on a hierarchy of regions and a hierarchy of production processes (Matsuhashi and Togashi 1988). Spatially, there is a strict, fourpartite hierarchy of the urban system: in terms of central management functions of large enterprises, Tokyo is considered the leader followed by Osaka, the regional cities (Sapporo, Sendai, Hiroshima, Fukuoka) and the prefectural capitals (Yada 1987). Relative to the technology-intensive industries in Japan this means that, besides other headquarter functions, the R&D headquarters, in particular, are located in the megalopolis of Tokyo-Osaka-Nagoya. From the hierarchy of production processes (R&D is very important; simple production and assembly are less important), it follows that less location-dependent manufacturing (meaning the kind for which there are several equally good locations) can be located in a region of low regional hierarchy. Translated to Japan's high-tech industries, this means that the headquarter functions and the R&D centres remain in the metropolitan regions where the required highly-qualified labour and R&D-intensive institutions are available and the necessary personal contacts are possible. Simpler production, on the other hand, (and also contractors depending on large enterprises) can be located in peripheral regions where the production factors 'labour' and 'space' of the required quality and quantity are available and where there is a good airport infrastructure. The latter is important because the transportation charges for many high-tech products, due to their special price-weight relation, have nearly no influence on the price per unit (Sargent 1987; Nishioka 1985).

A particularly good example for explaining spatial division of labour is the semi-conductor industry, because it is the most significant one in terms of quantity and the most dispersely distributed one of all eight high-tech industries in Japan (Nishioka and Takeuchi 1987). The R&D units are almost all located in the three metropolitan regions of Tokyo, Osaka and Nagoya. The technologically more advanced enterprises producing discrete and hybrid circuits are also rather oriented towards the centres. On the other hand, the businesses in the 'Silicon Island' of Kyushu concentrate on the assembly of prefabricated parts.

The fact that transport costs are irrelevant in the case of many high-tech products seems to render these industries footloose to a certain extent. Therefore, the question about the feasibility of decentralising technology-intensive industries was brought up in Japan and the answers from native regional scientists are under controversial discussion (Takeuchi 1991; Nishioka 1985). According to spatial division of labour, the likelihood of decentralisation quite obviously increases with decreasing complexity of production. The early R&D-intensive manufacturing stages require a location in the metropolitan regions; the later stage of intermediate assembly of prefabricated parts is labour- and space-intensive and can, therefore, be carried out at lower costs on the periphery; final assembly and distribution again take place in the centres (Nishioka 1985).

Spatial Division of Labour as an Explanation for the Dynamism in the Semi-Conductor Industry on Kyushu

Kuyushu owes its many high-tech establishments mainly to the semi-conductor industry. Consequently, the high-tech development in this region was primarily promoted by the locational demands of this industry between 1970 and 1990. These locational demands prove the plausibility of the theory on spatial division of labour in Japanese industry. The comparative advantages of Kyushu refer to only one high-tech industry and essentially to only one stage of manufacturing, namely the intermediate assembly of prefabricated parts.

During a relatively early stage in the product life-cycle of the semi-conductor industry – the labour-intensive manufacture of a product in increasing numbers – Kyushu possessed the necessary locational features: among them, first and foremost, the availability of reliable and flexible labour and, second, the much improved means of transportation (air and train traffic and, to a degree, road traffic), low-cost space and local economic policy (Sargent 1987). It was only due to these comparative advantages that Kyushu became a beneficiary of the decentralisation of high-tech industry with its positive effects on business.

The semi-conductor establishments on Kyushu are characterised by a high percentage of branch plants of companies with headquarters in Tokyo, by a lack of R&D-pursuing businesses and a lack of depth in their production line. Although, toward the end of the 1980s, some higher-grade production processes were also relocated to Kyushu (e.g. IC design centers of Sony in Nagasaki and Kokubu or Toshiba in Kitakyushu), over 90% of the semi-conductors produced there are not integrated into a regional production complex (development – production – intermediate assembly – final assembly, including suppliers), but are sent to Tokyo/Osaka for final assembly or exported directly (Matsubara 1992). These are not independent industrial districts according to Marshall (1919) nor technology districts according to Storper (1993), but externally controlled businesses and, thus, industrial regions. In 1986, in all prefectures, with the exception of Kumamoto, 70% of employees worked in businesses with headquarters in Tokyo or Osaka (Yamamoto 1991). The decentralisation of the high-tech industries into peripheral regions such as Kyushu, positive in principle, at a closer look turns out to constitute an extension of the metropolises' sphere of influence because – at least in the case of the semi-conductor industry on Kyushu – it caused a 'branch plant economy' or 'brainless economy' to develop, for which Kamino (1984) finds the term 'Silicon Colony' more appropriate than 'Silicon Island'. Intensive regional linkages between semi-conductor firms and local businesses would only be possible if the latter were also buyers of the products. Because of the underdeveloped local engineering industry, however, this is not the case, so that an endogenous development of technology-intensive businesses as a consequence of the settlement of the semi-conductor industry is not likely. All this seems to confirm the hypothesis of Glasmeier (1988) – at least with regard to the Japanese semi-conductor industry – '...that, in terms of linkage development, high-tech manufacturing establishments often behave no differently than other manufacturing branch plants' (p.276). Given these characteristics of the dominating high-tech branch and the system of spatial division of labour, the semi-conductor industry is not likely to become the long-term growth motor of regional economy.

The 'Technopolis' Strategy and its Impact on Regional Disparities and 'Silicon Island' Kyushu

In Japan there is a relatively strong economic, technological and political disparity between the three metropolitan regions of Tokyo, Nagoya and Osaka, on the one hand, and the peripheral parts of the country, especially in the north (Hokkaido, Tohoku) and south (Kyushu), on the other (Mutlu 1991). The disparate spatial structure is the cause as well as the consequence of numerous strategies of industrial policies following spatial goals and/or regional policies targeting on specific industries pursued since World War II. There have been both strategies explicitly aiming at strengthening the megalopolis of Tokyo-Nagoya-Osaka and strategies whose goal it was – and still is – to reduce spatial disparities (Hayashi 1991; Flüchter 1990; Cheung 1991; Abe 1995). In the more recent past, the 'Technopolis' programme was at the centre of the regional and technology policy debate (e.g. Tatsuno 1986; Kawashima and Stöhr 1988; Glasmeier 1988; Sternberg 1995c; Stöhr and Pöninghaus 1992; Castells and Hall 1994 and Chapter 8 by Park in this volume). By building technology cities (26 so far) in mostly backward parts of the country, the 'Technopolis' programme tries to achieve the introduction of high-technology into the entire national economy *and* the reduction of economic disparities between the individual regions. According to the latest amendment of the skeleton planning of 1991, the objective is the improvement of the quality of regional industries and manufacturing establishments, the promotion of the settlement of high-tech businesses, the intensification of linkages between the regions (e.g. on Kyushu) and the marketing of the comparative advantages of the individual regions and 'Technopolis' (Kyushu Industrial Advancement Center 1995).

Those responsible for 'Technopolis' formulated specific goals for each zone regarding foundation rates of industrial enterprises in general, and of high-tech enterprises in particular, employment and population figures, industrial value-added and business turnover, all of which must be achieved by a certain deadline. Nevertheless, to the author's knowledge, there are to date no (published) studies of all, let alone individual, zones or samples of zones (Tsukahara 1994) regarding the effects of the 'Technopolis' strategy on the intended goal variables – although this data is continually being gathered. Part of the reason lies in the local political 'explosiveness' of the projects, whose success or failure in public opinion is considered to depend directly on the prefectural governor advocating the respective project (Stöhr and Pöninghaus 1992). So far only two comprehensive empirical studies on the success of the technopolises in all areas have been published (Stöhr/Pöninghaus 1992; Sternberg 1995b) and they arrive at the following results:

- The gap in high-tech enterprise foundations (rate per year) in peripheral regions as compared to the core regions was reduced.

- The technopolises which are far away from the metropolises were able to improve their rate of business foundations significantly after the instrument had been implemented.
- Direct access to Shinkansen railroad stations has a favourable effect on business foundations in general and on high-tech foundations in particular.
- The absolute size of the zones and the high-tech foundations increases in inverse proportion to the distance from Tokyo.
- As regards increase in productivity, gross value-added and turnover (not, however, in employment), prefectures with technopolises showed a more favourable development than those without.
- All in all, the results confirm the hypothesis according to which technopolises which are close to metropolises are more likely to achieve the specific goals than are the more numerous locations on the periphery

There are other studies, using qualitative measure, stressing the positive aspects of individual 'Technopolis' elements, such as the reduced out-migration from rural areas (Tatsuno 1986) and the growing willingness to discuss regional development policy in the prefectures (Nishioka 1985). However, these predominantly positive assessments are contrasted with a good many sceptical investigations using primarily qualitative arguments (Glasmeier 1988). Doubt is expressed whether private R&D institutions can be decentralised (Takeuchi 1991) and local linkages can be created by means of high-tech industries (Glasmeier 1988). The majority of domestic and foreign observers arrive at the conclusion that the 'Technopolis' concept contributes to the spatial expansion of the 'Tokyo-Yokohama Mega-Technopolis' rather than to the decentralisation of high-tech industries (Castells and Hall 1994).

Six of the 26 technopolis zones are located on Kyushu (see Figure 9.3). They belong to the older zones (founded in 1984 and 1985). For each technopolis there is a skeleton plan containing exact target figures in terms of number of employees, inhabitants, industrial shipment and industrial value-added to be achieved by certain years. They can be compared to the values reached *de facto*.[1] The current skeleton plan, revised in 1991, is based on 1995 as the year of reference. The specific growth rates of the indicators mentioned above for each technopolis take account of the respective basic regional conditions.

1 This chapter is based on information from MITI Kyushu (1991) and the author's own research in the technopolis zones. The employment, turnover, value-added and productivity figures refer to manufacturing. The author wishes to thank the administrations of the technopolis zones and his Japanese colleagues (especially Prof. Matsubara and Prof. Yada of the Seinan Gakuin University Fukuoka, and Prof. Yamamoto of Hosei University, Tokyo) as well as Dr. Sang-Chul Park (University of Gothenburg) for their kind support.

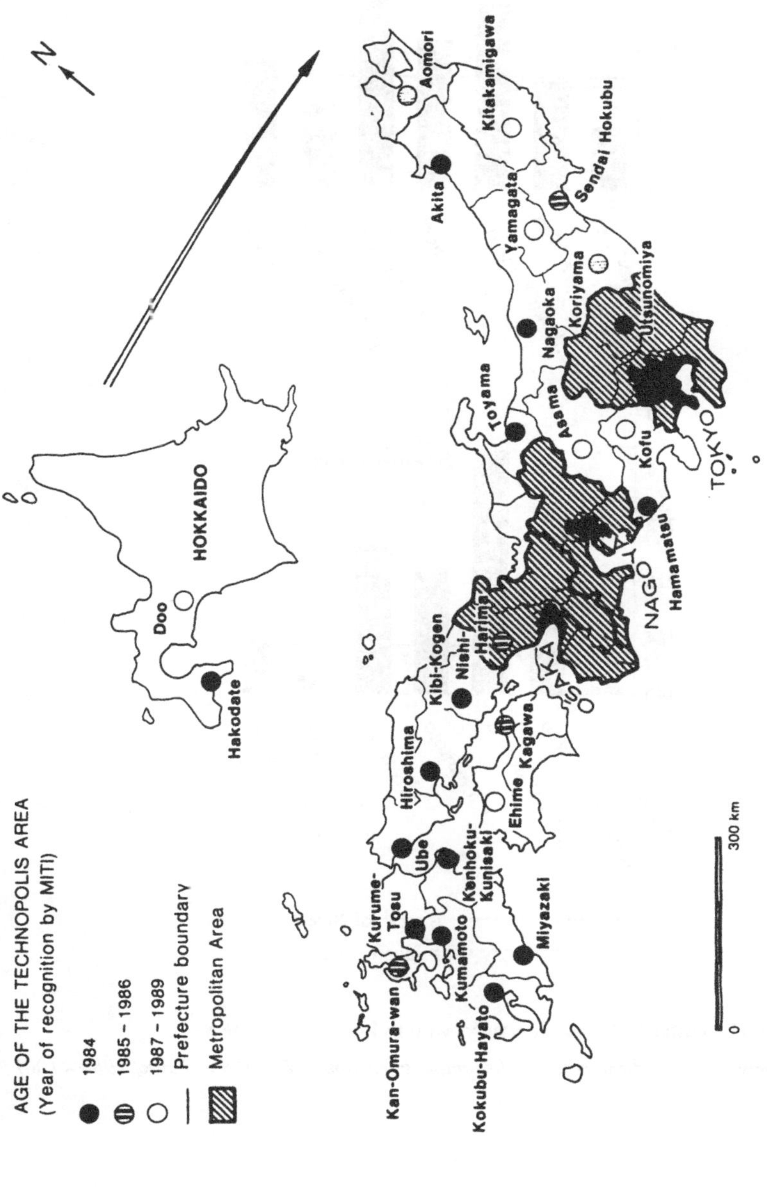

Figure 9.3 Regional distribution of Technopolis zones

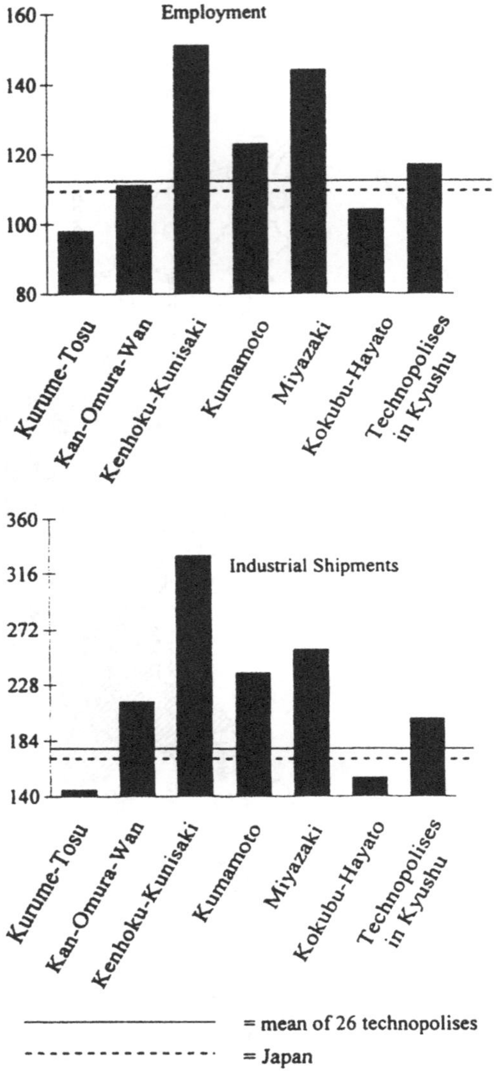

Source: MITI Kyushu 1991 and interviews with technopolis managers

Figure 9.4 Development indicators for Technopolises in Kyushu 1980–1991 (Index, 1980=100)

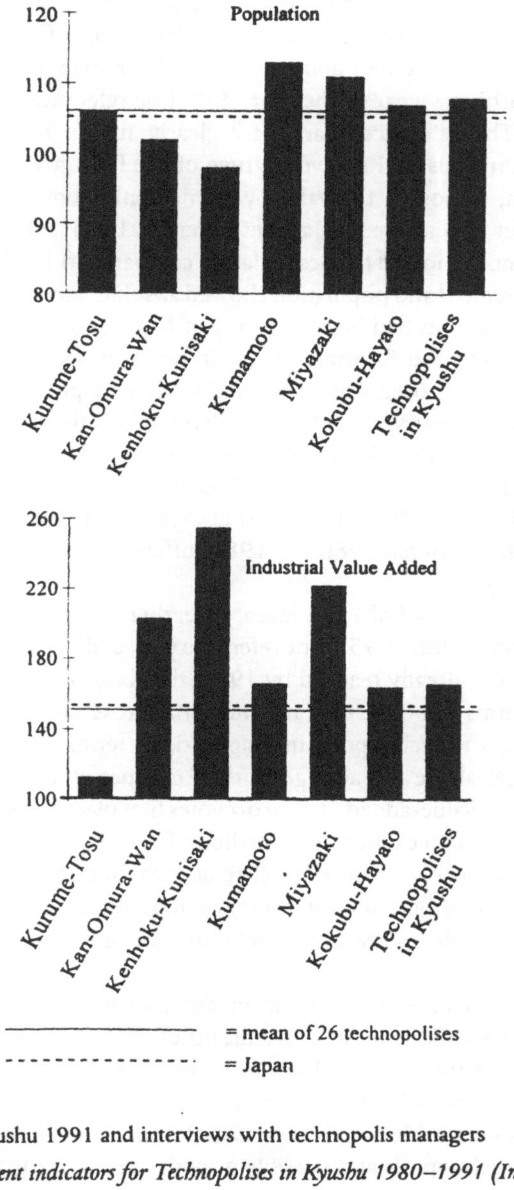

Source: MITI Kyushu 1991 and interviews with technopolis managers

Figure 9.4 Development indicators for Technopolises in Kyushu 1980–1991 (Index, 1980=100) (continued)

There are three methods for adequately evaluating both the absolute and relative success of the technopolis programme on Kyushu. The first method measures the specific success of the respective zone against the degree of goal achievement in percentages in terms of the four above-mentioned variables, the time of reference being either an arbitrary year or the year 1995 (the reference year of the current skeleton plan). The reference year 1989 clearly makes Kumamoto the most successful zone on Kyushu. Relative to three of the four goal variables (employment, population, turnover), the values were not only above the average of the Kyushu zones but also above the Japanese average. Until 1989, even absolutely speaking, Kumamoto showed the best balance; compared to 1980, the largest zone in terms of employment and population showed absolute increases relative to these two indicators (so far this had been true only of Miyazaki). The second best result was achieved by Kenhoku-Kunisaki which, in terms of goal achievement relative to employment and turnover, was above, and relative to, population barely below the national average. However, this technopolis was the only one on Kyushu having suffered absolute population losses since 1981. An overall negative balance results for Kurume-Tosu, Kan-Omura-wan and Kokubu-Hayato, which were unable to achieve any of the goal values and almost always stayed below the average value of the six zones and between 1980 and 1989 suffered absolute employment losses (Sternberg 1995b).

Later data covering part of 1991 reveal a clearly more favourable picture of the situation on Kyushu. With 1995 as the reference year, and relative to the percentage of goal achievement already reached by 1991, the six zones prove very successful – especially in terms of population and industrial gross value-added. Particularly Kan-Omura-wan, the technopolis in Nagasaki, in terms of all four indicators, shows percentages above the average of the 26 zones and, relative to industrial shipment and gross value-added, it even occupies first place. However, these results must be interpreted with caution. The method of working with a location-specific goal value is questionable because the criteria according to which the goals were set (and already modified several times) cannot be objectively verified. The possibility that, for political reasons, goal figures are kept low intentionally cannot be excluded.

Another method uses the increase in the indices of the above-mentioned indicators between 1980 and 1991 as a means of identifying growth processes of varying intensity in the technopolises. Although not all technopolises are of the same age, the differences between them appear to be rather small. However, this method is likewise unable to prove a causal connection between the time of foundation of a technopolis and possible growth rates. It turns out that:

- the technopolises on Kyushu, in terms of all four indicators, took a more favourable development than the other 20 zones, on average, and Japanese national economy as a whole
- the technopolises of Kumamoto and Miyazaki, in terms of all four indicators, surpass the average of all Japanese zones and also – contrary

to the otherwise more dynamic technopolis of Oita – show a far-above-average population development

- the 26 zones, on average, revealed a slightly higher growth than the overall Japanese national economy. (Figure 9.4)

The conclusion can be drawn that those zones which are relatively successful are those whose prefectures had a strong high-tech intensity of manufacturing establishments and employees even before possible technopolis effects had any influence (Table 9.1, p.163). The figures do not indicate a decrease in inter-regional high-tech disparities due to the existence of a technopolis.

The growth goals pursued with the Technopolis programme can only be reached if the number of endogenous or relocated businesses in the zones increases; above all, of course, these businesses have to be technology-oriented. Therefore, the third evaluation method compares the average annual number of business establishments (and/or high-tech establishments) in the technopolises before and after the opening of the respective zone. This comparison 'before and after' shows that among the businesses established after the opening of a technopolis in the zones of Kyushu, high-tech enterprises are relatively more strongly represented than in the 26 Japanese technopolis zones together. A slight trend toward the decentralisation of new establishments in general, and of high-tech establishments in the technopolis zones on Kyushu, is evident. As regards the total number of new establishments, the zones on Kyushu were, until 1989, less successful than their counterparts in other parts of the country.

In summary, relative to all four indicators, almost all the technopolises on Kyushu have undergone a more dynamic development than the other 20 zones, although they started at a very low level. Also, measured in terms of the goal figures for 1991 and 1995, the six Kyushu zones show favourable results. However, there is cause for scepticism. First, the relative progress in terms of dynamic indicators does not change the fact that, by absolute criteria, both Kyushu as a whole and its individual technopolises lag far behind most of the rest of the country. Second, it remains to be seen, whether an endogenous, lasting and R&D-oriented economic development beyond purely quantitative success will actually be set in motion by the Technopolis programme. The growth of the automotive industry – not necessarily a high-tech industry – deeply rooted in the region, does not give any indication to this effect. It seems realistic to assume that those parts of Kyushu which do not belong to the 'technopolis' zones will be far less successful in their efforts to establish new high-tech businesses.

Conclusions and Evaluation

Kyushu deserves the predicate 'Silicon Island' insofar as its relative number of high-tech employees does indeed far exceed the national average. However, this technology intensity is almost exclusively owing to one industrial branch, the semi-conductor industry. On the other hand, the term 'Silicon Colony' also applies

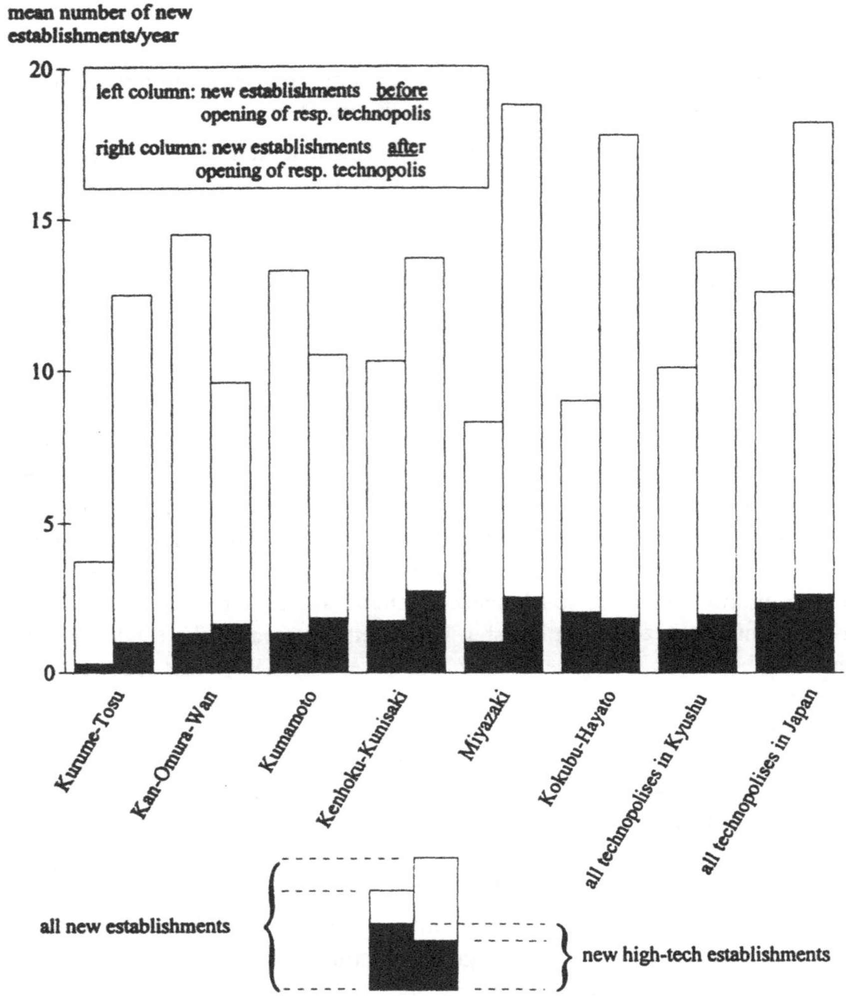

Source: MITI Kyushu 1991 and own interviews with technopolis managers

Figure 9.5 Mean number of new establishments/year of Technopolises on Kyushu before and after the official establishment of the respective Technopolis

because the high-tech businesses on Kyushu, in terms of function, heavily depend on the megalopolis of Tokyo-Osaka-Nagoya. The comparative advantages of Kyushu continue to consist of relatively low-cost (in terms of Japan, not all of south-east Asia) and well-trained labour. The locational conditions for the corresponding phases of the production process of high-tech industries are also, but not only, favourable here. Therefore, a high density of high-tech industries and strong external control do not constitute a fundamental contradiction. In terms of

Kyushu this means that so far it has not been possible to create an endogenous development dynamism with strong intra-regional linkages and an innovative milieu (the generally recognised cause and consequence of regional high-tech processes). These characteristics of a technopole (Simmie, Cohen and Hart 1993) are lacking on Kyushu. The regional and technology policy of the central government has taken advantage of this situation by integrating the comparative advantages of the regions into its strategies in favour of national competitiveness and large enterprises. Although, in the recent past, the six technopolis zones on Kyushu have displayed more dynamism than the other 20 technopolises have on average, one has to be careful not to expect too much. The Technopolis programme will probably not be able to significantly improve intra-regional linkages between the high-tech firms (which would be a precondition for a technopole), nor is it likely to contribute to a marked decrease in spatial disparities between the megalopolis of Tokyo-Nagoya-Osaka and the periphery, including Kyushu. Concerning 'Technopolis', we must consider that, up to the mid-nineties, almost every technopolis zone still critically depended on major public support in financial, political and organisational matters, which came mainly from central government (Steffensen 1995). The local parties are still too weak, not only in Kyushu, to initiate a shift from exogenous to endogenous development. That is one reason why local cities and regions, up to now, do not yet play the important role postulated in the Fourth Comprehensive National Development Plan (Edgington 1989; Abe 1995) as well as in the 'Technopolis' concept. Thus, the example of Kyushu shows that the association of 'Silicon…' with high-tech is not altogether incorrect; it is misleading, however, because high-tech does not necessarily mean high-level, crisis-proof, well-paying jobs nor does it mean endogenous and dynamic regional development.

References

Abe, H. (1995) *New Direction of Regional Development Planning in Japan.* Paper presented at the Regional Studies Association European Conference 'Regional Futures – Past and Present, East and West.' Gothenburg, Sweden 6–9 May.

Castells, M. and Hall, P. (1994) *Technopoles of the World. The Making of 21st Century Industrial Complexes.* London, New York.

Cheung, C. (1991) 'Regional innovation strategies and information society: a review of government initiatives in Japan.' *Asian Geographer 10*, 1, 39–61.

Edgington, D. (1989) 'New strategies for technology development in Japanese cities and regions.' *Town Planning Review 60*, 1, 1–27.

Flüchter, W. (1990) 'Japan: Die Landesentwicklung im Spannungsfeld zwischen Zentralisierung und Dezentralisierung (Japan: Regional Development between Centralization and Decentralization).' *Geographische Rundschau 42*, 4, 182–194.

Glasmeier, A.K. (1988) 'The Japanese technopolis programme: high-tech development strategy or industrial policy in disguise?' *Journal of Urban and Regional Research*, 268–283.

Hayashi, K. (1991) 'High-technology strategies and regional restructuring.' *International Journal of Political Economy 21*, 3, 70–89.

Kamino, T. (1984) 'Kuni no Sangyo Seisaku o Chiho Shudo de Okanau Mujun (Contradictions inherent in the Execution of National Industrial Policy through Provincial Initiatives).' *Asahi Jaanaru*, 3rd February.

Kawashima, T. and Stöhr, W. (1988) 'Decentralized technology policy: the case of Japan.' *Environment and Planning C Government and Policy 6*, 427–439.

Japan Industrial Location Center (ed) (1990) *Technopolis Policy of Japan.* Tokyo: Japan Industrial Location Centre.

Kyushu Industrial Advancement Center (1995) *Encounter Kyushu. Statistics, Facts and Figures.* Fukuoka: Kyusha Industrial Advancement Center.

Marshall, A. (1919) *Industry and Trade.* London: Macmillan.

Matsubara, H. (1992) 'The Japanese semi-conductor industry and regional development: the case of "Silicon island" Kyushu.' *The Economic Review of Seinan Gakuin University 27*, 1, 43–65.

Matsuhashi, K. and Togashi, K. (1988) 'Locational dynamics and spatial structures in the Japanese manufacturing industries: a review on the Japanese industrial restructuring process of leading sectors.' *Geographical Review of Japan 61*, (Ser.B), 174–189.

Ministry of International Trade and Industry (MITI) (1992) *1991 Establishment Census of Japan.* Tokyo: MITI.

Ministry of International Trade and Industry Kyushu (MITI) (1991) *Technopolises in Kyushu.* Fukuoka: MITI.

Mutlu, S. (1991) 'Regional disparities, industry and government policy in Japan.' *Development and Change 22*, 3, 547–586.

Nishioka, H. (1985) 'High technology industry: location, regional development and international trade Frictions.' *Aoyama Journal of Economics 36*, 2, 295–339.

Nishioka, H. (1991) 'Some aspects of Japanese history and economy – the land and people in Japan: Part 2.' *The Aoyama Journal of Economics 18*, 2, 19–51.

Nishioka, H. and Takeuchi, A. (1987) 'The development of high technology industry in Japan.' In M. Breheny and R. McQuaid (eds) *The Development of High Technology Industries.* London, New York, Sydney: Croom Helm.

Porter, M.E. (1990) *Competitive Advantage of Nations.* New York: Free Press.

Sargent, J. (1987) 'Industrial location in Japan with special reference to the semi-conductor industry.' *Geographical Journal 153*, 1, 72–85.

Saxenian, A. (1994) *Regional Advantage.* Cambridge, Mass: Harvard University Press.

Scott, A. (1988) *New Industrial Spaces. Flexible Production Organization and Regional Development in North America and Western Europe.* London: Pion.

Simmie, J., Cohen, J. and Hart, D. (eds) (1993) *Technopole Planning in Britain, Ireland and France: The Planned Regional Acceleration of Innovation.* London: Borttett School of Planning, UCL.

Steffenson, S.K. (1995) '"Techno" Obscuring "Polis" – a decade of public technopolis planning in Japan.' *Japan Forum 7*, 1, 67–93.

Sternberg, R. (1995a) 'Altindustrieregionen und Technologiepolitik auf nationaler und regionaler Maßstabsebene – die Beispiele Greater Boston/USA und Kyushu/Japan (Old Industrialised Regions and National and Regional Technology Policy – the Cases of Greater Boston/USA and Kyushu/Japan.' In D. Barsch and H. Karrasch (eds) *Umbau alter*

Industrieregionen. Verhandlungsband des 49. Deutschen Geographentages in Bochum (Vol. 1). Stuttgart: Steiner.

Sternberg, R. (1995b) 'Supporting peripheral economies or industrial policy in favour of national growth? An empirically based analysis of goal achievement of the Japanese "Technopolis" program.' *Environment and Planning C: Government and Policy 13*, 4, 425–440.

Sternberg, R. (1995c) *Technologiepolitik und High-Tech Regionen – ein internationaler Vergleich (Technology Policy and High-Tech Regions – An International Comparison).* Münster, Hamburg: Lit.

Stöhr, W. and Pöninghaus, R. (1992) 'Towards a data-based evaluation of the Japanese technopolis policy – the effect of new technological and organizational infrastructure on urban and regional development.' *Regional Studies 26*, 7, 605–618.

Storper, M. (1993) 'Regional "worlds" of production: learning and innovation in the technology districts of France, Italy and the USA.' *Regional Studies 27*, 5, 433–455.

Takeuchi, A. (1991) Impulses for Reorganization of Industrial Systems in Japan. Report of Researches, Nippon Institute of Technology 21, 2, 1–14.

Takeuchi, A. (1994) 'Locational dynamics of industry in the Tokyo metropolitan region.' *Report of Researches, Nippon Institute of Technology 3*, 3/4, 9–34.

Tamura, H. (1992) *Restructuring of Industry in the Kitakyushu Area, Japan.* Paper presented at the 7th Conference of German-Japanese Geographers in Heidelberg and Duisburg, 16–31 August.

Tatsuno, S. (1986) *The Technopolis Strategy. Japan, High Technology, and the Control of the 21st Century.* New York: Prentice Hall.

Tomozawa, K. (1992) 'Recent technological innovation in the Japanese automotive industry and its spatial implications for the Kyushu-Yamaguchi area in south-western Japan.' *Science Reports of the Tohoku University, 7th Series (Geography) 42*, 1, 1–19.

Tsukahara, H. (1994) 'A study of evaluating the technopolis policy from five indices.' *Annals of the Japan Association of Economic Geographers 40*, 3, 56–64.

Yada, T. (1987) 'On the regional structure of the Japanese economy.' *Keizaigaku Kenkyu 52*, 5, 41–55.

Yamamoto, K. (1991) 'Spatial patterns of control and dependency in Japanese corporations.' *Journal of International Economic Studies 5*, 57–77.

PART IV

Technology Transfer

CHAPTER 10

After Technopoles
Diffused Strategies for Innovation and Technology Transfer

Nic Komninos

Introduction

Science and technology parks (technopoles) are recognised as important institutions and infrastructures for industrial innovation and technology transfer, providing an interface between universities, R&D and production activities. However, after two waves of technopolitan development in Europe (1969–73 and 1983–93), major disadvantages in this technology transfer mechanism have become apparent. The chapter starts with a critique of the innovation and technology transfer concepts which are associated with technopoles and science parks, characterised by highly-localised and low-institutionalised technology transfer. The second section comments on some recent developments in the technology transfer policy of the European Commission, including the Strategic Programme for Innovation and Technology Transfer (SPRINT), the Regional Innovation and Technology Transfer Strategies scheme (RITTS), the Regional Technology Plans Pilot-Action (RTP) and the Fourth Framework Programme for R&D. Based on these experiences, the final sections discuss new tools for research-industry linkages and technology transfer which diffuse innovations through network structures and decentralised institutions. The recognition of the effectiveness of such tools and instruments for innovation support is gradually changing the nature and direction of technopolitan development.

Technopoles and Innovation Support

More than 25 years of technopolitan development in Europe have contributed to the creation of a phenomenon with substantial dimensions and expectations. The first phase of science and technology park development occurred at the beginning of the 1970s (1969–73), and it was rather experimental. It concerned a small number of cases, the science parks of Cambridge and Heriot-Watt in Britain, Haasrode in Belgium and Sophia Antipolis in France. These pilot projects appeared as spontaneous initiatives of the universities and private economic groups and searched for focal links between the universities and industries (Monck *et al.* 1988; Muller 1985). A second phase started at the beginning of the 1980s, during which the phenomenon received quite important dimensions with the creation of more than 100 parks in all European countries. These parks are connected to the wider political and economic framework of productive restructuring, the disintegration of productive activities, the rise of small businesses and the new demands for R&D, innovation and producer services. Since then, the co-ordinated efforts of the public and private sectors for new types of economic activity based on research and technology sustain the science park movement as an important institution supporting innovation and industrial competitiviness.

Science and technology parks vary significantly from one part of Europe to another. In the UK, Germany, Holland, Belgium and Greece a model of small 'incubator-led' parks predominates. These parks support new technology-based firms, on the levels of production, product development and finance. On the contrary, in France and Spain technology parks (technopoles) are larger and seek to change the entire local productive system, where they are located, through the attraction of large high-tech companies and multi-national R&D departments. This 'attraction-led' model acts as a catalyser in order to appeal to, and house, innovative firms in particular areas. In both cases, science and technology parks appear as zones/clusters of innovation and co-operation among R&D, industry and education. Networking and technology exchange link research institutions, innovative companies, new start-ups and supplier firms.

In most of the technopoles created during this period, the same components are more or less present (see Dunford 1992; Komninos 1992):

- a university-production co-operation, which creates a technology and innovation environment open to firms
- an infrastructure which transfers technology and business services to SMEs or larger firms
- a number of innovative firms, which creates a pole for innovation capable of diffusing technology and know-how to the wider productive system around the park.

Innovation and technology transfer activities constitute the nucleus of science parks; around it is set the population of the parks, consisting of R&D institutions, small and larger firms, infrastructure and supporting services. Technology transfer is realised through different forms of institutional agreements and management practices, including (see Dalton 1992; Komninos 1992):

- Agreements between firms and the universities. Generally they assure (1) the opening up of university infrastructures and research facilities to companies, (2)the flow of information and expertise through common projects, personnel transfer and day to day contacts, and (3) the support of the new ventures of scientists who wish to exploit commercially their research.

- Finance for start-ups of technology-based firms. In the absence of affluent seed capital financing, many parks have themselves organised seed and venture capital funds. The purpose is to support new start-ups, which constitute an important channel for technology transfer and innovation.
- Networking between firms. Strategic alliances in the fields of producer-supplier relationships, marketing-diffusion relationships, common R&D or product design projects and new joint ventures, resolve the usual difficulties that small firms face in production and marketing.
- Housing of innovative firms. An important part of the global activity of science parks is to provide new types of buildings and spaces characterised by flexibility and high quality. Such infrastructures attract the well-qualified employees of high-tech firms and promote parks as centres of innovative activity and business excellence.

Various claims are frequently made for the benefits and positive effects of science parks, including new firm formation, encouraging university-industry links, new employment and high-technology enterprise. However, in many cases the technological dimension and the technology transfer mechanisms of the parks were proved inefficient, technology transfer was neglected while other forms of entrepreneurial activity were favoured (see Massey, Quintas and Wield 1992; Massey and Wield 1992). Three situations, inherent to the constituting concepts of science parks, have contributed to such distortion.

First is the emphasis on the property dimension of the parks. Science parks were conceived as:

> a property based initiative which has formal operational links with a university or other Higher Education Institution as major centre of research; it is designed to encourage the formation and growth of knowledge based businesses and other organisations normally resident on site; and it has a management function which is actively engaged in the transfer of technology and business skills to the organisations on site (Dalton 1987, p.i)

A great deal of the science park management is related to property; to sell land, to build and to fill up incubators. In larger parks, where the stake of property is more important, property management caused a neglect of technology resources and job creation. As J. Hennebery (1992) points-out, one might be tempted to evaluate the impact of science parks in narrow cost-per-job terms by comparing the scale of public investment with the number of jobs created in 'academic' businesses.

A second reason is the emphasis on marketing and image strategies developed by the management in order to attract tenants. It was observed that an important motif for the location of firms was the high-tech image of the parks, which attracts tenants independently from the real technological potential (Monck *et al.* 1988). In many cases, marketing and promotion strategies prevailed, although the local business environment was very poor in innovative activities and the potential for the attraction of tenants very low (Technopolis International 1992, Komninos 1993).

A third reason is the low institutional links of parks with higher education and research institutions. Relatively few science parks have been developed by universities and many parks have been developed without functional relationship to academic and research institutions. This restrains considerably the supply of technologies and innovation services they are supposed to provide. An assessment of science parks role in the diffusion of technological knowledge (see Van Dierdonck *et al.* 1991; Van Dierdonck and Huysman 1992) notes the moderate technological environment that many parks offer.

These orientations of science parks towards physical accommodation and the attraction of innovative firms have diverted their function and tend to transform them from technology-supporting mechanisms to property-intensive developments. The spatial and marketing issues prevailed, while technology transfer, innovation and re-engineering of corporate practices were frequently neglected. Furthermore, the emphasis on space and physical infrastructures created a disproportion between the investments needed for the development of the parks and their real added-value in the innovation and modernisation processes.

A renewed interest for technology and innovation issues is needed if science and technology parks wish to achieve their policy objectives and remain important elements in the regional innovation infrastructures. Some new directions and solutions to this problem have been elaborated in the European Commission's innovation and technology transfer policies.

Technology Transfer, Innovation Support and EC Policies

During the 1980s and early 1990s, European Commission policies and programmes on technology transfer opened new ways of thinking about innovation support systems and shaped a number of infrastructures and services for such purpose. Most important have been contributions from the Strategic Programme for Innovation and Technology Transfer (SPRINT), the Regional Innovation and Technology Transfer Strategies and Infrastructures (RITTS), the pilot-action of Regional Technology Plans (TRP), and, recently, the Fourth Community Framework Programme for Research and Technological Development. These programmes have accumulated important experiences in technological co-operation and created generic tools for technology transfer and innovation diffusion.

SPRINT has been the main European Community programme for technology transfer during 1984–88 and 1989–93. It had three objectives:

- to facilitate the diffusion of new technologies to firms (support of specific projects for technology transfer, support for innovation financing by smaller firms, and inter-firm co-operation)
- to strengthen the European innovation and technology support services (support for science parks, innovation services, networks of technology and innovation specialists)
- to improve the awareness and understanding of innovation (creation of the European Innovation Monitoring System, support for the exchange of knowledge and experiences between the Member States).

Two major concepts were developed in the programme. First, the identification of *technology transfer routes* open to SMEs. Three basic technology transfer routes were identified: (1) the research to industry route, which can provide firms with sophisticated new knowledge, (2) the inter-firm technology transfer route, based on sub- and co-contracting relationships and (3) the technology licensing and related contractual forms of technology transfer (CEC 1994b). A second important concept was the building of *trans-European networks* for co-operation and application of new technologies in sectors and regions where they are yet to be utilised. These networks aimed at promoting inter-firm co-operation and helping SMEs from different countries to trade technology, carry out joint R&D, market complementary innovative products or to engage co-operation in the fields of technology and innovation. Since 1994, SPRINT was incorporated in the Fourth Framework Programme for R&D, as part of Third Activity devoted to the dissemination and exploitation of R&D results, technological development and demonstration.

Under the SPRINT Programme was developed the Consultancy Scheme for Regional Innovation and Technology Transfer Strategies and Infrastructures (RITTS). The scheme was aimed at regional governments and associated regional development organisations wishing to improve or change the focus of infrastructures and services for innovation and technology transfer. It covered a wide part of the Community, not just objective 1 or 2 regions, and it had a trans-national dimension in order to encourage the spread of best practices. Overall, 23 projects for regional innovation and technology transfer were supported. On the methodological level, the scheme was divided into three stages. The first stage was concerned with drawing up an inventory to define infrastructure support elements, business needs for R&D and types of possible public intervention; the second stage was concerned with the examination by a steering Committee of the strengths and weaknesses of the regional economy and the definition of a development plan; the third stage consisted of the implementation of the plan and follow-up mechanisms.

Complementary to RITTS were the Regional Technology Plans, a new pilot policy of the Commission to enhance the synergy between Technological Development Policy and Cohesion Policy. Regional Technology Plans were developed in the framework of the Structural Funds and were jointly managed by DG XVI (Regional Policy) and DG XIII (Telecommunications, Information Market and Exploitation of Research). They have launched in Leipzig-Halle-Dessau (Germany), Limburg (Netherlands), Lorraine (France) and Wales (UK). More recently four less-favoured regions have joined the programme: Abruzzo (Italy), Castilla y Leon (Spain), Kentriki Makedonia (Greece) and Norte (Portugal). Typical deliverables of a RTP include: the definition of a plan for technological development based on the consensus of the main actors in the public and private sector, which is to be implemented through the second Community Support Framework, Community initiatives and other investments from the public and the private sector; the organisation of a system for continuous monitoring and evaluation of technology issues and the needs of regional firms; the participation of the involved region in the network of Community's regions developing RTPs and building competitive advantages on technology and innovation. The management of each programme is based upon two local bodies: a Steering Committee and a Management Unit. The Steering Committee, composed of representatives from the public and private sectors, the Universities and other research institutions, oversees the whole operation and guarantees a regional consensus among the actors involved. The Management Unit assures the day-to-day work of the programme, launches the necessary studies and supports, scientifically and technically, the orientations from the Steering Committee (CEC 1994a; Landabasso 1995).

Major objectives of both the TRPs and RITTS (actually included in the INNOVATION Programme) are to encourage the endogenous technological development of the European regions, to improve the capability of local and regional actors to design policies which correspond to the real needs of the productive sector and the strengths of the local scientific community, and to support local consensus among the public authorities, the private sector and the universities about the character of technological development of the region.

An important push to technology transfer actually came from the Fourth R&D Framework Programme, which defines a number of priorities on the levels of technologies and the diffusion of innovation (CEC 1993b). The programme is divided in four activities which concern specific RTD projects, international co-operation, dissemination of RTD results and training and mobility of researchers. Major objectives shaping the programme are: the promotion of technologies having wide applications into a large number of industrial activities, and the promotion of dissemination and diffusion of R&D results.

The first activity of the programme, which covers 87.3% of the budget, is concerned with the development of generic technologies. Major applications include: information and communications technologies (28.2% of the activity's budget), energy technologies (18.5%), industrial technologies (16%), bio-medicine

Table 10.1 Technology orientations in the first activity of the fourth R&D framework programme

1. Information and Communication Technologies
   - Information technologies
   - Advanced communication technologies
   - Telematic applications of common interest
   - Technologies for integrated information and communication systems

2. Industrial Technologies
   - Design, engineering, production systems and human management
   - Materials and material-related technologies
   - Advanced propulsion technologies
   - Standards, measurement and testing

3. Environment
   - Natural environment, environmental quality and global change
   - Environment-related technologies
   - Earth observation and the application of space technologies

4. Life Sciences and Technologies
   - Biotechnology
   - Biomedicine and health
   - Applications of life science

5. Energy
   - Clean and efficient energy technologies
   - Nuclear safety
   - Controlled thermonuclear fusion

6. Research for a European Transport Policy
   - Strategic research for a multi-nodal trans-European network
   - Optimisation of networks

7. Socio-economic Research
   - Evaluation of science and technology policy options
   - Research on education and training
   - Research on integration in Europe and social exclusion phenomena

Source: Eurotechnology, No 41, January 1994

and biotechnologies (13.1%) and environmental protection technologies (9%). These choices reflect the technological needs of EU industries and services and link directly the Fourth Framework Programme to the European policy for industrial competitiviness (CEC 1993b).

However, the main challenge for translating RTD results into industrial innovation lies on dissemination and technology transfer. The programme includes three separate activities concerning the diffusion of RTD:

- the Second Activity for international co-operation, with 4% of the programme's budget
- the Third Activity which builds upon VALUE and SPRINT experiences and concerns the dissemination and exploitation of research results, with 2.5% of the budget
- the Fourth Activity for the stimulation of training and mobility of researchers, with 6.2% of the budget (CEC 1993b).

New Interfaces Between R&D and Production

The concepts and experiences acquired in the framework of SPRINT, RITTS, RTPs and the Fourth R&D Programme renew the discussion on technology transfer, research and industry co-operation and innovation support systems. These programmes have contributed in the development of new tools for the diffusion of R&D. Typical schemes are the observatories, the technology co-operation networks, the structures for the provision of advanced technological services and the institutions for innovation financing. Such tools may be incorporated in various technology transfer institutions to extend and change the character of the technology transfer infrastructure – science and technology parks included. In all cases, the central issue is the direct connection of innovation support initiatives to corporate strategies, the technology needs of SMEs and the competitiveness of local and regional productive systems.

Observatories and Technology Information

The key issue in technological development is to raise the awareness of the agents involved in productive practices and policy-making on R&D, innovation, technological capabilities and solutions. Different structures may be useful to this purpose: observatories, information centres, telematic networks, data-bases, etc.

Such information mechanisms rely on a double interface:

- A structure for the selection of information on technology issues that interest companies and, usually, SMEs; it covers both the supply of technological services by institutions of research and brokerage agencies as well as the demand for technologies and specific technological services.

- A structure for the diffusion of information to producers, public administration and the research community, covering a wide range of issues related to the specific problems of SMEs and the policies and programmes for technological development. This may include formal and informal procedures of communication, meetings and other forms of information exchange. Furthermore, the development of multi-media technologies and powerful bases for data storage and process permit the building of user-friendly interfaces for on-line communication and information.

Two good examples are EIMS and patent information. The European Innovation Monitoring System (EIMS) searches to establish a knowledge base and to develop research capabilities on innovation. It encourages the exchange of knowledge and experience between the Member States and the European Commission concerning innovation policies and innovation support measures through the development of a network linking experts and research teams performing applied innovation research and surveys at a European level, the systematic diffusion of results, studies and surveys performed in the EU and the establishment of a permanent Community-wide data-collection for monitoring the innovation capabilities and performance of industries and regions.

Complementary are the data bases on patent information. About a million patent documents are published every year in the 100 nations that have signed the Paris Convention for the Protection of Intellectual Property of 1883, which is the cornerstone of the modern patent system. Patent information may have a double role. On the one hand to provide inside information on existing competition, markets that might be exploited, starting points for R&D and informed reviews to the state-of-the-art in specific technologies. On the other hand it may provide solutions to specific problems and save development costs from duplication of research (see Derwent 1986).

Technology Transfer Networks

Technology transfer networks may be considered a major instrument for technology transfer. Technology networks are built upon supplier-producer relations, regional agglomerations of firms, international strategic alliances in new technologies, consortia for technological co-operation and specialists for long-term co-operation. The concept of network refers to a decomposable system in which the system is more than the sum of its interacting components; in other words there is synergy and multiplication effects from the interaction of the networks' members (see Cooke and Morgan 1991; Freeman 1990). Major issues for the technology networks concern focus, membership and the services (see also, Bianchi and Bellini 1991; De Bresson and Amesse 1991).

Concerning the focus, technology networks can be built on three different bases: to be focused on a sector-focus, on a technology, or on a combination of both.

- Sector-focused networks are very common and the narrow specialisation of industrial branches makes co-operation easier across wide geographic areas. Such networks may include firms, sectoral technical organisations and specialised research institutions.
- Technology-focused networks tend to be closer to the state-of-the-art in the area of expertise, but the lack of natural affiliation to particular industries makes the partnership more difficult. A useful base for technology-focused networks may be the Fourth Community R&D Framework Programme and the technological components of the first activity since it offers a typology of generic technologies and provides links to working groups, experts, research teams and industrial applications.
- Mixed-focus networks combine the benefits of the previous types, the facility for industrial application and awareness of the state-of-the-art of specific technologies.

For the membership, two different forms of technology co-operation may be distinguished: technology exchange, where technologies pass from one member of the network to another, and technology exploitation, where technological knowledge developed by research teams of the network is transferred to firms. So, with respect to the objectives for technological co-operation, the networks may include, firms, private consultants (including technology brokers, management consultants, consulting engineers, industrial property consultants, patent attorneys), research and technical organisations (contract research organisations and sectoral technical centres) and public and non-profit organisations, such as regional development organisations, chambers of commerce and industry.

Technology networks are able to offer a wide portfolio of services. Apart from information transfer, key services include technology transfer, skills transfer and specialist support (see Table 10.2).

The choices made on the focus, the membership, and the services, set the initial framework for the global management of the network: the internal co-operation and alliance, the leadership, the procedures and rules of management, the conflict and under-performance resolve, the creation of new markets based on synergy and the long term perspectives.

Provision of Advanced Technological Services

A wide range of technology applications and services is increasingly demanded by SMEs which cannot provide them internally, and which are not easily available on the market (see Britton 1989). Such services concern primary production and

Table 10.2 Services provided by technology networks

Technology transfer	*Skills transfer*	*Specialist support*
Technology brokerage	Training and education	Financial advise, market research
Licensing in, licensing out	Recruitment, Skill search	Technology application and management
R&D and technology audits	University – industry liaison	Demonstration
Research for products and processes	Location of R&D resources	Product evaluation, patent intellectual property advise

Source: CEC 1994b

research services, like certification, product development, multi-media applications, CAD-CAM applications, software and computational tools, and various types of laboratory analysis – including destructive and non-destructive quality analysis, chemical analysis, laser and optoelectronics applications, mineral exploration analysis and hydrogeology surveys and analysis.

Most demanded are the services related to:

- Quality certification, which is increasingly important for the competitive presence of products in the European and international markets
- Business services such as the technological evaluation of new products and firms (analysis of technology-based inventions, analysis of technical feasibility, market analysis), the protection of intellectual property (patents, model protection, registration of trademark, licensing, royalties), the design of development plans (business plan preparation, build and testing of prototypes, user questionnaires and product improvement) and marketing (choice of marketing route, establish a new venture/ find a partner licensing, design of marketing strategy)
- Multi-media and telecommunication services for product promotion and consulting for the suitability of multi-media for different applications and platforms. On the other hand, small systems of distributed informatics ensure that texts and voices may travel along the telephone lines and links may be established to data banks, to specific services available on a national and international scale and to local agencies through the installation of electronic mail services.

These services are linked to technologies which have become very important in contemporary production and product and management technologies like automation, quality control, energy saving, environmental protection and information technologies for business purposes. They have a direct impact on upgrading the technological level of productive processes necessary to improve competitiveness.

They are provided by different types of organisations, mainly in the public and semi-public sector, whose portfolio of services is concentrated upon SMEs' technology needs.

Innovation Financing

The creation of tools for innovation financing stems from the need to encourage new business start-ups. Technology-based and innovative SMEs have been misunderstood by bankers and financiers. Their characteristics differ markedly from those of more traditional businesses and this has led to a serious gap between businesses and financing institutions (CEC 1995).

When it comes to providing start-up and initial expansion financing for small-scale projects, the more conventional financing tools are ill-suited to companies' needs or are only partially able to satisfy them. There are, however, some tools specifically designed towards providing equity capital, which increases the chances of long-term survival. Their formation include (see CEC 1993a):

- To create the fund on an equity basis, probably in co-operation with existing financial institutions or banks. Alternatively, it is possible to create a separate finance line into an established venture capital fund.
- To set an appraising unit for innovative projects. What is needed is a methodology for evaluating the risks and the technical feasibility of proposals with respect to the market environment in which the new products will operate. Such appraisal may also consider the management, marketing and technical skills of the business and make use of external technology and marketing experts.
- To inform SMEs on the financial capabilities of the scheme and the comparative advantages *vis-à-vis* traditional financing tools.
- To design exit routes, always in co-operation with the financial institution or bank involved, for withdrawal from the individual project (financial route, industrial route, selling of the stake to the investee company, etc.).

Experiences and know-how on the functioning of such funds have been accumulated in the European Seed Capital Fund Network (ESCFN). It is a pilot scheme of the European Commission having, as it's overall objective, the fostering enterprise creation (and employment) in the Community by strengthening the financing opportunities available to new businesses. ESCFN supports a number of newly-created funds – provided that they agree to make their investment in start-up or early stage businesses. Some funds, if located in specific areas, may also benefit from a contribution to the funds available for investment.

Post-Technopolitan Command Centres

The experiences acquired by the technology transfer and innovation programmes of the EC, and the tools designed in this framework, show alternative routes which enrich and transform science and technology parks.

Science and technology parks have been started by universities wishing to valorise their research through the proximity and provision of physical accommodation to technology-based firms. The activity of high-tech firms, as well as of the start-ups in new industrial activities, have produced needs for new types of spaces – the chief characteristics of which were flexibility, quality and information technology infrastructure. In turn, the agglomeration of high-tech companies in science parks created poles of technological capability and diffused innovation around their location.

However, the spread of new effective tools for technology transfer, based on networks institutions and services, questions the established character of technopolitan development. The novel feature of these tools is that they operate without property or spatially-polarised dimensions. Instead, they are based on institutions, networks and information technology infrastructure. A new post-technopolitan profile is emerging, in which the functions of science and technology parks are diffused in many parts of the local productive system and the spatial aspects of technopoles become less important.

Indicative of such trends is that more and more technology transfer initiatives, based on technology networks, observatories and centres for advanced technological services, are developed out of science and technology parks. This multiplication of technology transfer initiatives involves a great number of social actors at the local, regional, national, sector and associative levels. The consciousness of the role of innovation in the defence of jobs and income has increased the demand for technology intermediaries to be included in local and regional development programmes. On the other hand, new tools for technology transfer and information diffusion are created by many universities (industrial liaison offices, career advisory units, technology information centres) without any formal reference to technopolitan structures. They take the form of networks and institutions supporting the university-industry co-operation and are placed under the usual university administration and decision-making.

From the multiplication and spread of non-spatial tools for university-industry co-operation, new technopolitan designs are emerging (see Komninos, Mercier and Tosi 1995). Already the technology transfer environment has deepened (see Miege 1992), leaving the isolated technology park as a memory at the outskirts of the city, as the unique technology intermediary. In the new innovation environment, composed by many players and dispersed initiatives, the real issue is the establishment of networks and 'command centres' in order to assure co-ordination, focus for the various initiatives and to avoid duplication and waste of effort. What is needed is co-ordination and guidance of the different agencies providing technology and markets information, technology inter-mediation, advanced technology services and innovation financing.

The nodal, property, and marketing-led technopoles are transforming and inserting into a multi-centre and multi-level research-production interface. The strengths of this post-technopolitan interface lay in the multiplicity of the tools for technology transfer it encloses, which do not rely on heavy and costly infrastructure or on the size of cluster of innovative firms. Networks and information channels do not presuppose the spatial proximity of the participating members and open their linking capacity over large geographical scales. Technology transfer institutions, technology networks and appropriate information technology infrastructure are appropriate answers to inefficiencies of the established technopolitan agglomerations and their built-in spatial constraints.

References

Bianchi, P. and Bellini, N. (1991) 'Public policies for local networks of innovators.' *Research Policy 20*, 487–497.

De Bresson, C. and Amesse, F. (1991) 'Networks of innovators: a review and introduction to the issue.' *Research Policy 20*, 363–379.

Britton, J. (1989) 'Innovation policies for small firms.' *Regional Studies 23*, 2, 167–173.

CEC (1993a) *Methodological Definition and Feasibility Study for Setting-up a Local Venture Capital Company.* CGF Enterprises, DG XVI.

CEC (1993b) 'Fourth R&D framework programme budget.' *I&T Magazine*, DG XIII, Winter 1993.

CEC (1994a) *Regional Technology Plans Guide Book*, 2nd Edition, DG XVI/DG XIII.

CEC (1994b) *Good Practice in Managing Transnational Technology Transfer Networks. Ten Years of Experience in the SPRINT Programme.* DG XIII and Coopers-Lybrand.

CEC (1995) *Capital d' Amorcage*, DG XIII/B3.

Cooke, P. and Morgan, K. (1991) 'The network paradigm.' Conference on Undefended Cities and Regions Facing the New European Order, Lemnos, August.

Dalton, I. (1987) 'Forward.' In Peat Marwick McLintock *Science Parks and the Growth of Technology-Based Enterprises.* Cardiff: UKSPA.

Dalton, I. (1992) 'Science parks: a mechanism for technology transfer.' *Topos – Urban and Regional Studies Revue 5*, 55–70.

Derwent (1986) *The Derwent Guide to Patents.* London: Derwent Publications.

Dunford, M. (1992) 'Technopoles: research, innovation and skills in comparative perspective.' *Topos – Urban and Regional Studies Revue 5*, 29–54.

Freeman, C. (1990) *Networks of Innovators.* International Workshop on Networks of Innovators, Montreal, May 1990.

Hennebery, J. (1992) 'Science parks: a property-based initiative for urban regeneration.' *Local Economy*, 326–335.

Komninos, N. (1992) 'Science parks in Europe: flexible production, disintegration and technology transfer.' In M. Dunford and G. Kafkalas (eds) *Spatial Implications of Competition and Regulation in the New Europe.* London: Belhaven.

Komninos, N. (1993) *Technopoles and Development Strategies in Europe.* Athens: Gutenberg.

Komninos, N., Mercier, D. and Tosi, A. (1995) *Science Park of Technical University of Crete.* Report to CEC, DG XIII.

Landabasso, M. (1995) *The Promotion of Innovation in Regional Community Policy: Lessons and Proposals for a Regional Innovation Strategy.* International workshop on regional science and technology policy, Himeji, 13–16 February 1995.

Massey, D., Quintas, P. and Wield, D. (1992) *High-Tech Fantasies: Science Parks in Society, Science and Space.* London: Routledge, Chapman and Hall.

Massey D. and Wield, D. (1992) 'Science parks: a concept in science, society, and "space".' *Environment and Planning D: Society and Space 10,* 411–422.

Miege, R. (1992) *Les parcs scientifiques et la Communauti Europienne.* Conference on Science Parks, Luxembourg, 5–6 May, 1992.

Monck, C., Quintas, P., Porter, R., Storey, D. and Wynarczyk, P. (1988) *Science Parks and the Growth of High Technology Firms.* London: Croom Helm.

Muller, A. (1985) 'Les mutants de Sophia Antipolis.' *Technopolis,* Paris, Autrement, No 74.

Technopolis International (1992) *Archipel Europe: Islands of Innovation,* No 8, June 1992.

Van Dierdonck, R., Debackere, K. and Rappa, M. (1991) 'An assessment of science parks: Towards a better understanding of their role in the diffusion of technological knowledge.' *R&D Management 21,* 2, 109–123.

Van Dierdonck, R. and Huysman, F. (1992) *Science Parks, their Evolution and their Evaluation.* Conference on Science Parks, Luxembourg, 5–6 May, 1992.

CHAPTER 11

Local Economic Development Strategies and Information and Communication Technologies

Keith Tanner and David Gibbs

Introduction

Information and communication technologies (ICTs), the convergence of telecommunications and computing are fast becoming the generic technologies of the modern world, with near-universal applicability for all sectors of manufacturing and services (Graham 1992c). It has been suggested that these technologies will form the prime basis for future economic development, with approximately 60 per cent of all employment within the European Union (EU) becoming either directly or indirectly dependent upon them by the year 2000 (Ungerer 1990). Furthermore, it has been argued that the future economic health of cities and regions will depend upon their active participation in such technological developments and that failure to participate will have important negative consequences (Graham 1992a). Regional and local administrations throughout the world have recognised the importance of ICTs for the economic and technological development of their areas. For example, some have attempted to play a strategic role in encouraging the development and uptake of ICTs by small- and medium-sized firms and in providing training for the local population (Gibbs 1993; Graham 1993). However, little is known about either the opportunities or the problems faced by local administrations in developing ICT strategies. There are only a few examples of policy implementation on the ground and policy development, in Great Britain at least, is occurring with no effective co-ordination (Graham 1991; 1992b). Moreover, much of the evidence for local policy developments is largely based upon anecdotal and secondary source evidence (see, for example, Graham 1994; Hepworth 1992). This chapter is an attempt to remedy this situation by providing an analysis of local authority-led ICT policy developments in Great Britain. Given the largely secondary source evidence for local policy makers' involvement in ICT policies, the research on which this chapter is based was intended to discover the extent of ICT policy developments and the form and aims of local ICT policies through direct contact with local authority officers.

ICTs and Local Economic Development

It has frequently been argued that advanced economies have entered an age where knowledge has become the basis of economic development and prosperity (Gillespie *et al.* 1989; Hepworth 1990). This knowledge takes the form of information, which has become a commodity to be exchanged, sold, transferred and used. As such, the application of information has transformed the economic base of advanced economies, so that some commentators have argued that we are entering the 'information economy' (Newton 1992). While definitions of what constitutes the 'information economy' are subject to debate (see, for example, Webster 1994), the term is often used as a paradigm to analyse the interrelated roles of information and communication technologies in the process of economic development (Goddard and Gillespie 1987; Hepworth 1990). It refers to three particular aspects of structural change in advanced economies:

- the growing contribution of information-related activities to wealth generation and employment
- the increasing centrality of new information technology, as a form of capital, in management, production and consumption processes
- higher levels of specialisation based upon the commodification of information, involving particularly the privatisation of public information and the externalisation of 'in-house' information services (Hepworth 1990).

In the UK, major changes have occurred in the information economy's main player, the telecommunications market, since the late 1970s. From consisting of a publicly-owned, public-service driven system oriented towards telephony, regulatory change and technological advance have led to a number of privately owned networks and the rapid development of cable, mobile and cellular services, with an increasing emphasis on non-voice services. The combination of progressive digitalisation of telecommunications networks and investment in new high-capacity or 'broadband' lines and switches are the current key technological developments in advanced telecommunications. The resulting systems facilitate restructuring of organisations in terms of their organisational, spatial and functional makeup. Liberalisation of the market has also been an important factor in ICT developments. In 1981 the UK government abolished the public monopoly of BT by privatising the company and allowing the establishment of a competitor – Mercury Communications. Legislation in the form of the 1991 White Paper *Competition and Choice: Telecommunications Policy for the 1990s* allowed for the introduction of more competition into this duopoly (Department of Trade and

Industry 1991). Further competitors have been established, for example Ionica was granted a licence in February 1993 to set up a new nationwide telephone system. Advanced data transmission and information services have developed rapidly, largely in response to the demands of the corporate sector. It is increasingly being proposed that the role of ICTs in economic development is now very much an active element where the provision of advanced services and data transmission facilities are crucial to organisational development, such that '"intelligent" data services, high-speed global information services, facsimile services, sophisticated private exchanges and local area networks are now common currency in the major business centres of the UK' (Graham and Dominy 1991, p.181).

In the UK, the liberalisation of telecommunications has led to a much greater emphasis on serving the corporate market and upon increasing profitability. At the same time, the major losers in the new scenarios are residential areas, voluntary organisations and small- and medium-sized firms (SMEs). One of the key issues in the future development of ICTs is the extent to which access to information and ICTs is distributed equitably, not only between groups of users but also over space. Thus it is argued that cities form the main locus of these rapid developments in ICTs so that the transactional structure of urban areas is being re-modelled by their integration into the network economy (Gottman 1983; Graham and Dominy 1991). The extent to which areas and cities are participating in the major changes introduced by ICTs will determine a major part of their economic fortunes (Gillespie and Williams 1988). However, ICT investment decisions and policy-making have occurred in the UK at the national level, with little urban or regional component. This lack of any nationally-co-ordinated policy to assess the spatial implications of these decisions, and the perceived importance of ICTs, has led a number of local administrations to attempt to intervene positively in their own local economies.

This process is part of a more widespread engagement with the local economy by local administrations (Eisenschitz and Gough 1993). The large-scale deindustrialisation of the 1980s has forced administrations to become more 'entrepreneurial' and to develop new policy avenues rather than equate economic development with improvement of the built infrastructure (Harvey 1989). One such avenue is the development of the 'information city', in order to gain some influence over the development of, and interconnection with, new ICT infrastructures. At one level this concern is to remain competitive *vis-à-vis* other locations in an increasingly global economy, but there is also a concern with the social and economic impact of ICTs at the level of communities. Thus local policy needs to address uneven ICT developments not only at the spatial scale but also at the intra-urban level, to ensure that groups of individuals and firms within the city are not information-disadvantaged (Castells 1989). Local economies are increasingly integrated into world-wide networks of markets and industries through the globalisation processes aided by ICT developments. As national structures decline in importance (while still setting the regulatory context) the importance of the 'global-local nexus' increases (Robins 1989). Cities and regions, therefore, need to evaluate their role within the emergent global system and devise policies to position themselves in these new structures (Knight 1989). Thus, the ways in which this ICT system develops unevenly between and within cities and regions is becoming a crucial determinant of local economic fortunes (Gibbs and Tanner 1994).

Differential levels of investment in ICT networks are creating varying levels of comparative advantage at the international scale. Cities, for example, are becoming increasingly located within an international urban hierarchy. In this hierarchy, certain cities will function as 'international information capitals' and other cities will serve as 'regional information hubs' linked to the main capitals but with a more limited spatial field (Moss 1987). For urban administrations then, the crucial point is to either become a 'hub' or to ensure that adequate linkage to the network exists. For all cities and regions a good ICT infrastructure may be essential to attract inward investment and skilled labour (Cornford and Gillespie 1992). Without this linkage, peripheral cities, and regions in particular, become vulnerable to the processes of increasing concentration and centralisation of economic activity in core areas. However, the problem of linkage is compounded by the liberalisation of telecommunications. Telecommunications companies may not find it profitable to invest in certain areas or, at least, are not investing in the leading-edge technologies in these areas.

The adoption of ICTs is also very uneven at the organisational level. The large corporate sector is much more likely to be taking advantage of the technology than SMEs (Luthje 1993). In the UK, the large corporate sector has benefitted the most from deregulation as BT and Mercury compete for their business. By contrast, SMEs are unable to take advantage of the reduced costs and market reach engendered by ICTs (Goddard and Gillespie 1987). This may give rise to a 'dual economic base' within local areas, with large firms connected to the global economy, while SMEs are trapped in a restricted and impoverished local economy (Cornford and Gillespie 1994). Given the increasing reliance on this sector for indigenous growth at the local level, it is perhaps not surprising that local policy aims to act as a counterweight to these developments. Promoting networking amongst SMEs has been seen as a priority for economic development, allowing SMEs to gain access to national and international networks and interact with larger firms. The advent of the Single European Market and the need for SMEs, especially in more peripheral localities, to extend their market reach has also given a boost to ICT strategies.

Access to ICT developments also has a social dimension wherein access to information, skills and employment is also uneven, both across space and across social groups. Already information-disadvantaged groups (women, ethnic minorities, disabled people) and areas (the inner cities) are becoming increasingly detached from ICT networks and increased polarisation may result (Gibbs 1993). Some commentators have argued that this polarisation is an inevitable concomitant

– the information city is by necessity a dual city (Castells 1989). Again, the situation in the UK has been exacerbated by the reliance upon market forces to deliver ICT provision.

Attempts to mitigate the marginalisation involved at the spatial, organisational and social scales is therefore rapidly becoming one of the major policy challenges for local administrations. In particular the lack of an effective regional or urban ICT policy in the UK has forced administrations to 'develop new scenarios for future urban economic development upon which private-public attempts at regeneration can be founded' (Graham and Dominy 1991, p.186), despite the fact that they have no statutory role in such developments. The benefits of developing an ICT strategy are thought to include providing a comparative advantage over other cities and areas, some local control over the local information economy and a reduction in social and spatial inequality in access (Graham and Dominy 1991). However, the newness of these policy initiatives means that they are fraught with uncertainty. Which technologies and facilities should be provided and promoted? Are the costs of inertia greater than those induced by choosing the wrong technologies? Given local administrations' lack of in-house expertise, to embark upon ICT strategies can be risky indeed.

ICT developments can be seen as a 'double-edged sword' which not only provide inequalities and policy problems, but also give rise to opportunities for policy development (Gillespie *et al.* 1989). While infrastructural provision and network interconnection are vital, the greatest opportunities are said to lie in developing and promoting innovative uses and applications of ICT services (Ducatel 1994). Such services have been sub-divided by Miles and Thomas (1990) into three types: informational, communication and transactional services. Informational services, as might be expected, provide information mainly in the form of remotely-accessible databases. Communication services include voice mail and electronic mail, while transactional services include Electronic Data Interchange (EDI), remote banking and reservation services (Miles and Thomas 1990; Ducatel 1994). In terms of local authority initiatives, research in the late 1980s revealed that most local authority involvement was with informational initiatives – whereby the authority developed its own databases to provide or sell information to individuals or firms (Hepworth, Dominy and Graham 1989). Examples of local authority policy initiatives which involve transactional or communication services are more limited. Ducatel (1994) outlines the five main forms of involvement in ICT policy which have developed to date:

(1) providing telecommunications infrastructure upgrading and development as part of physical redevelopment.

(2) sponsoring telecommunications infrastructure investment as part of economic development.

(3) partnership investment in the telecommunications infrastructure to increase local authority leverage in the provision of services.

(4) the direct promotion of ICT services, providing all three types of ICT service.

(5) encouraging dialogue on ICTs between private sector firms.

However, Ducatel (1994, p.66) goes on to state that 'the number of cases in which local government initiatives are more than statements of intent filling space in strategy documents is quite limited. There seems to be only a few case studies of direct involvement in telematics projects by local development agencies'.

Methodology

In consequence, the starting point for the research upon which the next section of this chapter is based was this lack of empirical evidence for the development of local-level ICT policies. Some literature does exist which considers the development of such local strategies in a broad sense (see, for example, Goddard and Gillespie 1987; Hepworth 1989; Gillespie and Robins 1991; Graham 1992b, 1992c and 1993; Devins and Hughes 1993). However, as Graham (1994, p.426) states 'very little is known at this stage about these policies, aside from the catalogue of broadly descriptive information about individual policy initiatives'. There are few examples, and even less empirical evidence, of 'ICT locality studies' – that is studies which attempt to understand the ways in which ICTs are being used and promoted across the sectors of a local economy through local policy. In the absence of such empirical evidence and case study detail, much current literature on ICTs, economic restructuring and local policy responses is unsubstantiated, general and theoretical (Dabinett and Graham 1994).

The initial stages of the research have been aimed at investigating the use of ICTs and the development of policy within specific local economies, as well as discovering the extent to which policies and initiatives are in existence. The major published British study is the work of Graham and Dominy (1991) which investigated local telecommunications policies as part of the PICT programme at the University of Newcastle. This study also investigated the pattern and extent of local authority ICT policies but took a specifically urban focus through a survey of 93 British urban local authorities, of which 34 responded to the survey. 'Big' in this study, in the form of population size, was effectively equated with 'most advanced' strategies. No comprehensive national review was undertaken and the study focused on urban initiatives only. In order to investigate the extent and form of policy development in Great Britain, it was decided to conduct a postal survey of all 514 British local authorities at metropolitan, district and county council levels in July 1993. Using the Municipal Yearbook, questionnaires were sent to the economic development department, or its equivalent, in all 514 authorities. A total of 202 completed questionnaires were returned, a response rate of 39.3 per cent.

ICT Policies and Local Authorities in Great Britain: the Survey Evidence

Current and Planned Policy

The survey results revealed that ICT initiatives *are* increasingly forming part of the economic development activities of British local authorities. Of the 202 respondents, 13.5 per cent currently have an economic development strategy that includes a policy on ICTs and a further 21.3 per cent planned to incorporate ICTs into their strategy in the near future. In total then, 35 per cent of British local authorities in the survey had a current or planned ICT policy as part of their economic development measures. Obviously the existence of policy statements does not indicate action and this is a point that is returned to when looking at initiatives in operation. In terms of local authority type, the evidence would suggest that the literature to date has perhaps placed an undue emphasis on urban policy developments. A higher proportion of county councils (31.3 per cent) had a current ICT policy compared with district councils (10.6 per cent) and metropolitan boroughs (7.1 per cent). A high proportion of county councils also planned to introduce ICT policies, as did the metropolitan boroughs (see Table 11.1).

Table 11.1 Current and planned ICT policies by local authority type

	Current Policy (%)	*Planned Policy (%)*	*Total (%)*
Metropolitan Boroughs	7.1	39.3	46.4
District Councils	10.6	16.9	27.5
County Councils	31.3	25.0	56.3
Total – All Authorities	13.4	21.3	34.7

Source: Survey data

At the sub-national scale, Wales had the largest proportion of local authorities with a current policy (46.2 per cent), while the North-West had the largest proportion with a planned policy (45.0 per cent). Regions with little intention to include an ICT policy in their economic development strategies and with no current policy, were the South-East (78.6 per cent had neither a policy nor plans to introduce one) and East Anglia (85.2 per cent). Some explanation for this may lie in the existing provision of ICT services. While a quarter of all respondents felt that their area had deficiencies in its telecommunications infrastructure compared to other areas, only 7.1 per cent of authorities in the South-East and 18.5 per cent in East Anglia perceived local deficiencies. By contrast, local infrastructural deficiencies were recorded in Wales (by 53.8 per cent of authorities), the East Midlands (38.5 per cent) and the North-West (35.0 per cent). At the level of local authority type, deficiencies were also more common amongst county councils than metropolitan and district councils – although the differences were much less marked than at the sub-national level.

Rationale for Policy Development

The main reasons for having developed or planned a local ICT policy for economic development were to help develop small firm networks and to encourage inward investment (see Table 11.2), with few differences by local authority type. However, despite the fact that ICT policies are being introduced by a substantial proportion of British local authorities, the importance of the local telecommunications infrastructure was seen as less important as a factor in economic development policy than other,more traditional factors (see Table 11.3). However, there may be a mismatch here as research quoted by Cornford and Gillespie (1994) suggests that, for the EU's leading firms, the quality of telecommunications infrastructure in an area is second only to ease of access to markets and customers as a locational factor. Given the stated importance of attracting inward investment (Table 11.2) as a policy rationale, the perceptions of local authority policy makers may therefore be lagging behind the needs of industrialists.

Table 11.2 Reasons for possessing or planning an ICT strategy in British local authorities

	Per cent of respondents
To aid SME networks	65.7
To attract inward investment	64.3
To aid local community development	62.9
To promote access to the Single European Market	44.3
To promote access to other international markets	34.3

(Multiple responses possible)
Source: Survey data

Table 11.3 Factors ranked as very important for economic development

	Per cent of respondents
Transport infrastructure	61.7
Cost of land and property	50.0
Image of the area	50.0
Cost of and availability of labour	33.2
Telecommunications infrastructure	16.6

(Multiple responses possible)
Source: Survey data

Local Authority Involvement in ICT Initiatives

In addition to indicating the level of current and planned policy development within their local authority, respondents were also asked to indicate whether there were any current or planned ICT initiatives within their local authority area with which the local authority has some input. An interesting finding which emerged was that a higher proportion of authorities (20.3 per cent) had some involvement with an existing ICT initiative than had existing policies (13.4 per cent). An additional 17.3 per cent had plans to develop an ICT initiative. This illustrates Sellgren's (1987) observation that local authorities do not always develop initiatives based on the prior establishment of policy. Such initiatives can develop in an *ad hoc* fashion, with justification for the policy occurring, if at all, at a later date. Indeed, anecdotal evidence from contact with local authority officers, during and subsequent to the survey, has suggested that ICT initiatives have often been led by the availability of funding, rather than any strategic objectives.

As with policy, the operation of such ICT initiatives was more in evidence in county councils (43.8 per cent with a current initiative) than in metropolitan boroughs (17.9 per cent) or district authorities (15.5 per cent). A quarter of both county councils and metropolitan boroughs had future planned ICT initiatives. At the sub-national scale, the leader in current ICT initiatives was Wales (53.8 per cent of respondents had a current ICT initiative), followed by the Northern region and the East Midlands (both 23.1 per cent).

Those authorities with current and planned ICT initiatives as a sole or joint venture were asked to give brief details of such initiatives, including aims and objectives, target groups and funding sources. Not all respondents answered this question. A total of 116 separate initiatives were identified, with details supplied in varying degrees of completeness, from 62 separate authorities. Some local authorities had more than one initiative in operation, with the extreme case of Manchester City Council having 36 initiatives in operation (31 per cent of the total). In the case of this data, it proved more difficult to observe strong regional or local authority type differences.

ICT Initiative Objectives

Many respondents had more than one objective for their initiatives. The three most important were: to provide access to information (41 initiatives), to provide training in ICTs (40 initiatives) and for business development (36 initiatives). Few respondents mentioned inward investment or market access, which suggests that the operation of initiatives is somewhat at odds with stated policy objectives (see Table 11.2, p.203). While local community development was also rarely mentioned explicitly, it was often part of the development of informational services. For example, initiatives included developing community information displays at library access points, providing on-line employment information for women returners to the work-force and information displays on HIV/AIDS. Other, more commercial, examples of information provision included the development of on-line data bases for independent film makers, providing information on environmental legislation to businesses and developing on-line company directories.

While the importance of information provision confirms Hepworth, Dominy and Graham's (1989) finding that much local authority ICT policy is in informational services, a large number of local authorities with business development objectives were concerned with developing communicational services or, more rarely, transactional services. Such business development objectives fell into two main categories: developing networked workspace for small firms, often in a 'telecottage' style development, and promoting networking between firms in the local economy. Only a handful of initiatives could be classed as transactional, with the main focus here upon promoting trading links with partners in overseas markets and developing EDI.

Training objectives were particularly seen as a means of equipping unemployed people and disadvantaged groups with ICT skills. For example, in the East Midlands two such schemes were aimed at redundant miners with the hope of transforming economies that had been devastated by pit closures. Several schemes were targeted at women, particularly women returning to the work-force.

Target Groups for ICT Initiatives

The most important target group for ICT initiatives was the private sector, which was mentioned in 79 cases. An important sub-set of this group were small- and medium-sized enterprises, which were the target of 47 initiatives. Other notable target groups included the local community (usually through public information provision schemes), disadvantaged groups (such as disabled people and ethnic minorities), and unemployed people. The latter two target groups were, not surprisingly, usually the target of ICT training initiatives.

Funding Sources for ICT Initiatives

The majority of these ICT initiatives were joint initiatives between the local authority and a variety of other bodies, including: Training and Enterprise Councils (or Local Enterprise Councils in Scotland), the National Computing Centre, telecommunications companies, Chambers of Commerce and voluntary organisations. Not all the partners necessarily contributed funding to the initiatives. The most important funding sources came from the local authority's own budget (55 cases) and European Union funds (38 cases) – including the European Regional Development Fund (ERDF), the European Social Fund (ESF), New Opportunities for Women (NOW) and RECHAR. Other sources of funding were from UK central government (24 cases) (mainly from the Urban Programme [now defunct] and City Challenge), Training and Enterprise Councils (15 cases), the private sector (13 cases) and the Rural Development Commission (9 cases). Few initiatives involved only one source of funding.

Barriers to Developing ICT Initiatives

Finally, all 202 respondents were asked to identify what they saw as the major obstacles to implementing and developing ICT initiatives for local economic development (see Table 11.4). It is evident that lack of finance dominates local authority reasons for non-involvement. However, some differences emerge at the sub-national level. While Wales, the East Midlands and the Northern region all ranked lack of finance as their major problem, an absence of qualified personnel was perceived as the major problem for Scottish authorities. Conversely, Welsh authorities cited neither lack of personnel nor the lack of political will as potential barriers. In East Anglia a higher than average number of authorities (33.3 per cent) mentioned a lack of political will as a barrier. The lack of a national co-ordinating body for ICT developments was an important factor in the East Midlands and Wales.

Table 11.4 Major barriers to implementing ICT initiatives

	Per cent of respondents
Lack of finance	63.7
Lack of a national co-ordinating body	18.8
Lack of qualified personnel internally	18.3
Lack of internal political will	15.8
Present internal organisation	10.4

(Multiple responses possible)
Source: Survey data

Conclusions

The growing recognition that ICTs pose both a threat and an opportunity to local economies has prompted a policy response from local administrations. The survey evidence presented here provides the first comprehensive overview of the local policy response by local administrations in Great Britain. The evidence reveals that ICT policies are fast moving up the local policy agenda: 35 per cent of the 202 local authorities in the survey have a developed or planned ICT policy and 38 per cent have a current or planned ICT initiative. The survey shows that the development of such policies and initiatives is most advanced in Britain's more peripheral areas, particularly the regions of Northern England and in Wales. Policy developments and initiatives are much less in evidence in Southern England, where the perception is that current infrastructures are adequate. Much of this evidence confirms the earlier work of Graham and Dominy (1991) that the 'distance shrinking' capabilities of ICTs are being used to try and overcome peripherality. However, it also suggests that the emphasis in the literature upon urban policy developments needs some adjustment. The most active policy and initiative developments were observed in county councils, which suggests that there is a need for more research into the aims and objectives of ICT policy in rural areas.

It is still rare, however, to see any analysis of exactly how such ICT strategies will have a directly beneficial outcome for the local economy. Strategies have developed in a piecemeal fashion and often without a clear idea of the economic development benefits they are supposed to achieve. In some cases there is a suspicion that initiatives are proceeding because funding is available, rather than because they fit with a strategy for the local economy (see Bovaird 1994). The main objectives of those local authority initiatives in operation were to: provide access to information, provide training (especially for unemployed people and disadvantaged groups), and to encourage business development (especially in the form of encouraging small-firm networking and networked workspace). However, adopting a supply-side approach and creating better infrastructural provision may not necessarily lead to increased economic activity. Policy makers cannot assume that the provision of ICTs is a sufficient condition for economic advance within a city or region. Establishing an infrastructure may be a precondition to development, but it does not automatically result in the provision of effective or relevant services to local businesses or other sectors of the community (Gillespie 1991).

In relation to SME networking, for example, the network between firms must exist before it can be mediated through ICTs and policy cannot create networks where none existed before simply through providing an ICT infrastructure (Melody 1991; Graham 1992c). The assumption that networking will occur (and the implicit assumption that some form of 'local production complex' or 'industrial district' will result) appears to underlie much local authority ICT policy directed to SMEs. The evidence is, however, that ICT usage is more likely to promote the 'processes of economic globalisation rather than supporting the re-emergence of neo-Marshallian industrial districts' (Dabinett and Graham 1994, p.616). In addition, even in Manchester (the authority with the largest number of ICT initiatives) SME involvement has been extremely limited (Gibbs 1992; Gibbs and Leach 1994). Small firm participation in ICT networks may be motivated by customer pressures rather than the desire or ability to interlink with other local small firms, which are more likely to be seen as competitors than collaborators (Graham 1994). The lesson for local authority policy makers here is that they cannot make assumptions about automatic use by firms in the local economy. There is a need for a much more pro-active approach which demonstrates the potential uses of the technologies to firms on an on-going basis (Henderson 1992). Even so, to have any impact, a whole set of innovation policies are needed – driven by integrated development institutions (Cooke and Morgan 1991). In relation to training initiatives, the assumption appears to be that training in ICT usage will improve an individual's chances of future employment. However, Dabinett and Graham's (1994) study of Sheffield suggests that firms are looking for potential workers with general communication skills, rather than technological expertise. Indeed, such ICT expertise was seen as a national labour market and a relatively unimportant issue at the local level. It can be argued that the types of ICT training being provided are not yet in sufficient demand in the labour market, although these projects undoubtedly have social benefits in raising individuals' confidence and personal skills base. In some cases this lack of hard evidence for demand is related to poor monitoring within the initiatives themselves. For example, training for women in the Manchester 'Electronic Village Hall' project is said to have resulted in 20 per cent of trainees gaining employment, but there is no monitoring to evaluate whether these trainees have gained technology-related jobs (Basker 1994).

Finally, it is worth pointing out the fragmented nature of both policy and initiatives and the limited financial resources available to local authorities. Compared to the major ICT players in Britain (BT in particular, as well as Mercury and the growing involvement of the cable companies), local policy is likely to have only a marginal impact. Respondents to the survey indicated that the major constraint upon developing ICT initiatives is lack of finance (Table 11.4, p.206). The detail of local initiatives revealed that sources of funding frequently involve a complex mix of funding and partners from the local authority's own budget and other local agencies such as the Training and Enterprise Councils, the European Union and central government. Of particular note is the limited role of central government resources available to local authorities and the growing importance of European funding. However, local responses in concert with the European Commission are no substitute for a co-ordinated national response to the challenge of ICTs which takes account of the needs of particular areas, firms and groups of individuals. The current British government's unwillingness to countenance this, and reliance instead upon market forces and liberalisation to deliver ICT infrastructure and services, seems likely to exacerbate regional, urban, corporate and social inequalities at a time when ICT developments are becoming ever more important.

References

Basker, S. (1994) 'Trainee comparative needs analysis and evaluation of IT and telematics training at the Manchester EVHs.' *Report of New Opportunities for Women Project*, Centre for Employment Research, Manchester Metropolitan University.

Bovaird, T. (1994) 'Managing urban economic development: learning to change or the marketing of failure?' *Urban Studies 31*, 4/5, 573–603.

Castells, M. (1989) *The Informational City*. Oxford: Blackwell.

Cooke, P. and Morgan, K. (1991) 'The network paradigm: new departures in corporate and regional development.' *RIP Report No. 8*, University of Wales, Cardiff.

Cornford, J. and Gillespie, A. (1992) 'Cable systems and the geography of UK telecommunications.' *PICT Policy Research Paper 21*, ESRC, Swindon.

Dabinett, G. and Graham, S. (1994) 'Telematics and industrial change in Sheffield.' *Regional Studies 28*, 6, 605–617.

Devins, D. and Hughes, G. (1993) 'Local economic development: can IT work?' *Seminar paper*, Policy Research Unit, Leeds Business School, Leeds Metropolitan University.

Department of Trade and Industry (1991) *Competition and Choice: Telecommunications Policy for the 1990s, Cm 1461*. London: HMSO.

Ducatel K. (1994) 'Transactional telematics in the city.' *Local Government Studies 20*, 60–77.

Eisenschitz, A. and Gough, J. (1993) *The Politics of Local Economic Policy: The Problems and Possibilities of Local Initiative*. London: Macmillan.

Gibbs, D. (1992) 'The development of Manchester's telematics strategy.' *Manchester Geographer 13*, 47–59.

Gibbs, D. (1993) 'Telematics and urban economic development policies: time for caution?' *Telecommunications Policy 17*, 4, 250–256.

Gibbs, D. and Leach, B. (1994) 'Telematics in local economic development: the case of Manchester.' *Tijdschrift voor Economische en Sociale Geografie 85*, 3, 209–223.

Gibbs, D. and Tanner, K. (1994) 'Information and communication technologies, local economic policy and regulation theory.' *Spatial Policy Analysis Working Paper 27*, School of Geography, University of Manchester.

Gillespie, A (1991) 'Advanced communication networks, territorial integration and local development.' In R. Camagni (ed) *Innovation Networks: Spatial Perspectives*. London: Belhaven.

Gillespie, A. and Robins, K. (1991) 'Non-universal service? Political economy and communications geography.' In J. Brotchie, M. Batty, P. Hall and P. Newton (eds) *Cities of the 21st Century: New Technologies and Spatial Systems*. Harlow: Longman.

Gillespie, A. and Williams, H. (1988) 'Telecommunications and the reconstruction of regional comparative advantage.' *Environment and Planning A*, 20, 1, 311–321.

Gillespie, A., Goddard, J.B., Hepworth, M. and Williams, H. (1989) 'Information and communications technology and regional development: an information economy perspective.' *OECD STI Review*, April 5, OECD, Paris, 84–111.

Goddard, J. and Gillespie, A. (1987) 'Advanced telecommunications and regional economic development.' In B. Robson (ed) *Managing the City: Impacts of Urban Policy*. Beckenham: Croom Helm.

Gottman, J. (1983) *The Coming of the Transactional City*. Maryland: Maryland Institute for Urban Studies, University of Maryland.

Graham, S. (1991) 'Telecommunications and the local economy: some emerging policy issues.' *Local Economy 6*, 116–136.

Graham, S. (1992a) 'Electronic infrastructure and the city: some emerging municipal policy roles in the UK.' *Urban Studies 29*, 755–781.

Graham, S. (1992b) 'The role of cities in telecommunications development.' *Telecommunications Policy 16*, 3, 187–193.

Graham, S. (1992c) 'Ringing the changes: telecommunications, local economic restructuring and local policy innovation.' In M. Geddes and J. Benington (eds) *Restructuring the Local Economy*. Harlow: Longman.

Graham, S. (1993) *Networking the City: Telematics Services and the Urban Policy Challenge*. Paper to PICT National Conference 'European Dimensions in Information and Communications: Panacea or Pandora's Box?', Kenilworth.

Graham, S. (1994) 'Networking cities: telematics in urban policy – a critical review.' *International Journal of Urban and Regional Research 18*, 3, 416–432.

Graham, S. and Dominy, G. (1991) Planning for the information city: the UK case. *Progress in Planning 35*, 3, 169–248.

Harvey, D. (1989) *The Condition of Postmodernity*. Oxford: Blackwell.

Henderson, D. (1992) 'The use of telecommunications to regenerate a rural area.' *Proceedings of International Conference on Data Transmission – Advances in Modem and ISDN Technology and Applications*. Institution of Electrical Engineers, London.

Hepworth, M. (1989) *The Geography of the Information Economy*. London: Belhaven.

Hepworth, M. (1990) 'Planning for the information city: the challenge and response.' *Urban Studies 27*, 537–558.

Hepworth, M. (1992) 'Information services and local economic development in Organisation for Economic Cooperation and Development.' *Cities and New Technologies*. OECD, Paris.

Hepworth, M., Dominy, G. and Graham, S. (1989) 'Local authorities and the information economy in Great Britain.' *Newcastle Studies in the Information Economy, Working Paper No. 11*. Centre for Urban and Regional Development Studies, University of Newcastle-upon-Tyne.

Knight, R. (1989) 'City development and urbanisation: building the knowledge-based city.' In R. Knight and G. Gappert (eds) *Cities in a Global Society, Urban Affairs Annual Review, 35*. Newbury Park, Ca.: Sage.

Luthje, B. (1993) 'On the political economy of "Post-Fordist" telecommunications: the US experience.' *Capital and Class 51*, 81–117.

Melody, W. (1991) 'The information society: the transnational economic context and its implications.' In G. Sussman and J. Lent (eds) *Transnational Communications: Wiring the Third World*. London: Sage.

Miles, I. and Thomas, G. (1990) 'The development of new telematics services.' *OECD STI Review*, July, OECD, Paris.

Moss, M. (1987) 'Telecommunications and the economic development of cities.' In W. Dutton, J. Blumler and K. Kraemer (eds) *Wired Cities: Shaping the Future of Communications*. Washington, DC: Communications Library.

Newton, P. (1992) 'The new urban infrastructure: telecommunications and the urban economy.' *Urban Futures 5*, 54–75.

Robins, K. (1989) 'Global times.' *Marxism Today*, December, 20–27.

Sellgren, J. (1987) 'Local economic development and local initiatives in the mid-1980s.' *Local Government Studies 13*, 51–68.

Ungerer, H. (1990) *Telecommunications in Europe*. Brussels: Commission of the European Communities.

Webster, F. (1994) 'What information society?' *The Information Society 10*, 1–23.

CHAPTER 12

National Laboratories and Regional Development
Case Studies from the UK, France and Belgium

Helen Lawton Smith

Introduction

This chapter examines national laboratories in three European countries, the UK, France and Belgium, from two points of view – as incubators of new firms and as institutions which potentially have a role to play in generating or contributing to an innovative milieu. The literature on regional development strategies and public institutions has tended to focus more on industry and academic links (e.g. Segal Quince 1988; Lawton Smith 1991; Palfreyman 1989) rather than national laboratories but, as this chapter demonstrates, the latter can have an impact where other local conditions are favourable. The evidence is taken from a study of nine laboratories undertaken between April 1993 and May 1994. The sample includes atomic energy authority laboratories in each country.

The rest of the chapter is organised into three sections. The first discusses the conceptual background to the study, the second provides some historical context to the activities of specific institutions and the third analyses sample laboratories as incubators and as agencies in regional development taking the UK, Belgium and France in turn. Finally, some conclusions are drawn.

The Concept of Innovative Milieu

The term 'innovative milieu' has come to mean a particular type of local economic development. The concept has been defined by Camagni (1993, p.3) as 'a set of relationships occurring within a geographical area, which unify a production system, different actors, an industrial culture and a self-representation, cumulatively generating a localised dynamic process of collective learning' (see also Storper 1993). This combines place with the idea of innovation as a social process built on collective knowledge and co-operative effort and which flourishes where scientific, technical and market information is readily exchanged and practical interaction is frequent (Sayer and Walker 1992 p.155). The dynamics of information exchange and implied learning can be seen as a spatial integration of skills in the localised component of a more extensively constructed scientific labour market whereby expertise resident in different parts of the innovative milieu is used to enhance that in others. The ability of institutions, such as national laboratories, to contribute to the development of a milieu therefore relies on the absorptive capacity of local firms (Cohen and Levinthal 1990).

In spite of the numerous studies of innovative milieu (see, for example, Cooke 1993; Garnsey *et al.* 1994; Keeble 1990) and the interest in demonstrating that development has become a localised phenomenon (Ettlinger 1994), it has been demonstrated that only a few places exhibit the kind of transaction-intensive linkages which the concept encapsulates. Moreover, the problem of scale is generally not addressed. Comparisons between what are considered to be innovative milieu are therefore difficult. This is because scale means different things within and between different countries, due to such spatial dimensions as the political construction of spaces (administrative authority), the kinds of industrial concentrations (pre-dominance of industrial sectors) and the degree of concentration or dispersal of industrial activity within a region – all of which may affect the functioning of the local socio-economic system.

Role of State Intervention

The key difference between the UK, Belgium and France is that, in the UK, regional development strategies and the regional development functions of national laboratories are two separate concepts. The first relates to the absence in the UK of a regional tier of government. However, in Scotland and Wales, where for some time there has been a devolution of some regional power, (Cheshire *et al.* 1992) there are possibilities for targeted initiatives at the regional scale (see Huggins 1995, for example, on the South Wales technopole). The second relates to the 'missions' of individual laboratories, which encompass a range of activities operating primarily at national and international scales; this means that the welfare gains from externality gains from public investment in science and technology are not acquired locally.

It is not surprising, therefore, that the literature on the UK tends to neglect the role of the local public infrastructure when discussing the growth of small firms in the context of the development of innovative milieu. However, studies in Italy, Germany and the USA, find that intervention at the local level in supporting entrepreneurship has contributed to the uneven spatial distribution of places (see, for example, Bianchi and Gordiani 1993 on Italy; Cooke 1993 on Germany; Saxenian 1991 on Silicon Valley) while its absence is seen as an inhibiting factor in France (Longhi 1995 on Sophia-Antipolis).

Incubator Organisations

The concept of incubator organisations is used here as a means of identifying some of the dynamics which can be involved when institutions such as national laboratories are mandated to support entrepreneurial activities. The concept encompasses more than the mere act of gestation. Incubator organisations theoretically fulfil a variety of functions, particularly with regard to their influence on local labour markets. It has been argued that they affect the kinds of people hired and brought into an area (Cooper 1971, p.2), thereby contributing to the process of developing the skill mix within local labour markets, and provide breeding grounds for new entrepreneurs who then act as role models for other founders. In these ways they can generate local multiplier effects, both economic and in the creation of a collective image, thereby contributing to the shaping and re-shaping of the local industrial milieu through time (Lawton Smith 1991). In this scenario, therefore, they can be a source of new (generally small) firms, influence the cultural/industrial character of areas and contribute to the skill base of the region, increasing spatial and technical segmentation in local (and, consequently, more geographically extensive) labour markets. By these means they can be agents of regional development.

National Laboratories: Historical Context

National Laboratories are here defined as those research laboratories which operate under the aegis of public authorities, even though they may not be directly funded from the public purse. The important distinguishing criterion is that their function and operating constraints are determined directly by a central (or regional in the case of Belgium) government department (e.g. defence, energy or industry).

The history of national laboratories dates back to the end of World War II, a time when European and North American governments promoted large-scale R&D programmes in strategic areas such as energy, defence and space (Heim 1988). However, it was the commitment to nuclear energy, to which nearly every OECD member in the post-war period devoted substantial R&D resources, which 'marked the beginnings of big science and big technology' (OECD 1989, p.21). Space exploration became the other spectacular growth area of government-funded research in the post-war period. These two developments led an expansion in the government research sector, which continued with the creation of new laboratories geared to particular needs. For example, in 1947 the UK established the National Mechanical Engineering Laboratory, now the National Engineering Laboratory (NEL).

From the post-war period, laboratories in the case study countries (and others) became an integral part of national scientific research effort and their functions defined a characteristic of the national scientific system. For example, in the UK, although they have had different missions to universities, they have had some characteristics in common. Indeed many have, or have had, close relationships with the higher education sector, and have had important educative functions. Interaction has included funding research in universities, the supervision of graduate students and staff serving as visiting lecturers. In this environment, it has been common for laboratory personnel to move to senior appointments in universities. In France and Belgium, supervision of students and university lecturing also is a well established tradition (see Lawton Smith forthcoming (1997)).

In the mid-sixties, rather earlier than Belgium, the UK and France began to introduce measures to increase greater utilisation and exploitation of national scientific and technological resources through their use in industry. In the UK, moves towards commercialisation of research have resulted in a reduction in these educative functions while in France and Belgium this has been accompanied by an expansion of the range of activities which support training and technology transfer activities, in some cases at the local level. A characteristic of UK national laboratories is that they tend not to be integrated into local information carrying and generating networks. This is due to a complex set of factors including the organisation of 'big science' in the UK (Ergas 1993), the lack of a mission to be involved in regional development strategies and, in some cases, geographical isolation.

The UK

In recent years the UK has adopted a more radical approach to re-defining the role of national laboratories than other countries. It has introduced measures to increase 'economic efficiency', such as the creation of Executive Agencies under the 'Next Steps' Initiative which began in February 1988 and was designed to bring improvements in the quality and efficiency of government services. The UK is now in the process of moving some national laboratories into the private sector, rather than keeping them within the state sector as in France and Belgium.

The UK sample comprised three laboratories: the Harwell Laboratory, the largest laboratory of the United Kingdom Atomic Energy Authority (UKAEA), NEL and Defence Research Agency (DRA) Malvern. The UKAEA or AEA as it is now known, was divided into two on May 16 1994: UKAEA (government science) and AEA Technology (the commercial organisation). AEA Technology is in the process of being privatised; the process began in March 1995. Harwell has divisions of both parts of AEA.

Two of the three laboratories are in the southern half of the country, where most are located. NEL, located south of Glasgow at East Kilbride, is the only national laboratory in the north of the UK. Its location is an outcome of conflicts between the defence ministry, which wanted to keep research institutions in the southern part of the country, and distribution-of-industry policies of post-war Labour governments (Heim 1988, p.376).

Harwell

The Harwell laboratory is part of a complex of six government research establishments in rural south Oxfordshire which includes AEA Culham Laboratory, the Daresbury Rutherford Appleton Laboratory and the National Radiological Protection Board. At March 1987, these together employed 7,710 people. Despite this concentration of research activities (the highest in Europe), local economic development based on entrepreneurship and regional development has been extremely limited; only 8 out of over 200 Oxfordshire advanced technology firms were founded by former employees of these laboratories (Lawton Smith 1990) and only three came from nuclear energy establishments. Harwell had not, therefore, acted as an incubator of new entrepreneurial small firms. However, the structure of the AEA as a whole is already split into some 150 different commercial businesses. At the time of writing it was not known whether AEA Technology will be sold off intact, or whether fragmentation will occur, which may result in the absorption of some of the units into existing firms or the creation of new entities – a different form of new-firm formation.

Part of the explanation for the lack of an incubation role is contractual limitations. Under the terms of the 1977 Patent Act, employees working in research establishments are expected to produce work that is patentable as part of their employment. Therefore the intellectual property belongs to the institution and not to individual scientists and engineers. Another factor, suggested in a study in the late 1980s, was that scientists who were likely to become entrepreneurs would not work for Harwell, the staff in general being 'risk averse' (Lawton Smith 1990).

Although Harwell has no brief to contribute to local economic development, it has a wider technology transfer function. It operates a number of collaborative 'clubs' in which small and large firms are members. These are research, development or technology transfer activities carried out at a host organisation. Many of these are supported by the Department of Trade and Industry (DTI). The first club at Harwell was Heat Transfer and Fluid Flow (HTFS), formed in 1967. By October 1987, the DTI had funded 104 clubs of different kinds, with some 2,430 members. The UKAEA as a whole was host to 15 clubs (DTI 1988).

NEL

NEL has two major research areas: Flow, and Energy and Environment, and is the major testing centre for national standards for flow measurement. In October 1990 NEL became the ninth DTI Executive Agency. NEL's current primary aim is to achieve commercial viability in preparation for privatisation. To achieve that aim, its objectives include adopting a more commercial approach to business, to develop new business and make more efficient use of its site (NEL Annual Report 1993). This policy is designed to bring NEL more into contact with industry in Scotland.

In the past, NEL has not contributed to the local economic development, either as an incubator of spin-out firms or by transferring technology or its staff into local industry. NEL has traditionally had a low level of links with local engineering industry. These have further declined with the reduction in the traditional engineering base in Scotland. This area of Scotland cannot be considered an innovative milieu in spite of the concentration of electronics firms in the central lowlands, an area which has come to be called 'Silicon Glen' (see Dunford 1989). This concentration is based on inward investment by firms such as Motorola (USA) and NEC (Japan), rather than indigenous growth. Indeed, patterns of local sourcing by these firms indicate a dependent model rather than a self-sustaining development scenario (Turok 1993) – the latter being a characteristic of innovative milieux.

Therefore, both the decline of the engineering base and the growth of inward investment at the possible expense of indigenous entrepreneurship, have increased NEL's geographical isolation from most of its client base and have decreased the number of potential employers for former laboratory employees. This study revealed that very few people made redundant in the recent restructuring process found jobs in Scotland (Lawton Smith 1994). This also meant that, in its traditional role, there was limited potential for a regional technological role in creating local synergies based on a local match of interest.

However, the focus of NEL's activities has changed in the 1990s. It is the only one of the UK sample which is part of a strategy for regional development. In 1994, NEL's 70 acre site became a 'technology park' owned by Scottish Enterprise, as part of a large initiative to develop the area south of Glasgow. NEL is a tenant. The location of firms on the park with complementary interests to NEL is being encouraged. By early 1994, some 25 small firms were on-site. Indeed, spin-off is now being considered by NEL as part of its changing remit.

At another scale, NEL is active in generating technology transfer. For example, NEL runs the flow liaison club which meets regularly, moving between different industrial and academic locations. In 1995, a new club, the 'NEL Multiphase Flow Club' was formed. By April 1995, ten members – from the UK, USA and Norway – had signed up (Flow Tidings, No9, April 1995).

DRA Malvern

The Defence Research Agency was established as a Next Steps Agency in 1991 and has operated as a Trading Fund from 1 March 1993. It has remained in the ownership of the Secretary of State for Defence. DRA Malvern is the technology base for the DRA. Its two largest businesses are Electronics and Command Information Systems.

DRA's ability to act as an incubator, or contribute to a local innovative milieu, is limited by regulation and by geography. First, spin-off is limited for security reasons, because of the military focus of R&D. Second, its location is a major factor inhibiting its regional development role. The laboratory occupies a site in a scenic part of rural Worcestershire, an area which has none of the characteristics of an innovative milieu. A DRA-funded feasibility study of the potential for a science park close to DRA Malvern suggested that, as there are a large number of this type of property development within 50 miles, most firms locating on the site would have to relocate from elsewhere. Indigenous development would be slow because of the low formation rate in the local economy.

In sum, in the UK, opportunities for contributions to local development are limited by the local environment: each of the three laboratories are in locations which do not have the characteristics of innovative milieux. Therefore, in this sample, historical and geographical factors as well as an absence of institutional mechanisms inhibit local economic development. The laboratories are not incubators of new firms and the presence of a concentration of scientists and engineers does not appear to have any associated multiplier effects, either by employment of staff in other organisations or from mechanisms designed to promote skill transfer from the laboratories. The national and international impact of the technology programmes of these institutions is a different issue.

Belgium

The focus of analysis in this study is the Flanders Region. Two laboratories will be discussed in this section: the Inter-university Micro-electronics Centre (IMEC) and the Studiecentrum Voor Kernenergie (SCK), the national nuclear energy centre. The focus is mainly on IMEC because of its unique role in regional economic development. The SCK, however, has recently adopted a pro-active strategy towards entrepreneurship.

Belgium is a prime example of a small country which has devolved responsibilities for economic development towards regional institutions. Belgium consists of three Regions: Brussels-Capital, Wallonia and Flanders. In 1980, the Regions were endowed with precisely defined powers and effective bodies. During 1988/9 the Constitution was revised and special laws adopted governing the Communities and Regions, their financial resources, the Region of Brussels-Capital (which did not have any administrative or decision making autonomy) and the Court of Arbitration. Since that period, the central state has approximately 60% of the resources while the Communities and Regions receive about 40%. After 1993 the Dehaene government had succeeded in reforming the constitution and Belgium became a federal state with directly elected regional parliaments appointing their own regional governments (Mommen 1994).

The process of decentralisation in industrial policy began in the late 1950s. From 1959 the incumbent Liberal-Christian-Democratic government began to display a selective economic policy. A system of investment subsidies was established to channel industrial restructuring into new growth sectors and to promote adaptation to the Common Market system. This instrument was also intended to attract international productive capital, particularly US corporations seeking to benefit from the new European market. By 1987, overall about one-fifth of Belgian industry was under direct American control. By the mid-1980s, about 60% of industry in Flanders was controlled by foreign or mixed groups (Moulaert and Willekins 1987).

Post-1980, Flanders adopted an offensive innovation policy known as DIRV. This came into force in 1981 and was the first important initiative of the new Flemish government. At that time it was seen that multi-national companies in Belgium were doing well but that Belgian industries were in crisis. The new Flemish policy stressed the development and application of advanced technology in industry. Universities were to be important in regional development by working with industry, and firms would be given research subsidies. IMEC was established in 1984 by the Flanders government to introduce state-of-the-art R&D to the Region's firms. This was part of a comprehensive programme to promote education, research and applications in micro-electronics, and so improve the industrial tissue of Flanders – while reducing dependence on technology developed outside the region and encouraging inward investment of the R&D functions of foreign firms. At the same time, a number of structures came into being – notably the Flemish research and technological development agency, the (IWT) formed in 1991.

A brief description of IMEC's remit is given first. This is followed by a discussion of the laboratory's role of incubator and some information about, and comment on, its training functions.

IMEC

IMEC's partner universities are the Katholieke Universiteit Leuven (KUL), Ghent University (RUG) and the Vrije University Brussels (VUB). The main location is at Heverlee in Leuven, adjacent to KUL, from where it originated. When the laboratory was founded, IMEC employed 100 staff and by 1994 this had grown to over 400. Some 70% of staff are scientists and 61% have university degrees. The age profile is very young: the average age is about 30. This balance is preserved by the appointment of the overwhelming majority of R&D staff to short-term contracts, for example for the duration of their PhD studies. Turnover of staff is around 15% each year. IMEC's annual budget is about 1.3 Billion BF (IMEC Scientific Report 1992). Its income is derived mainly from the Flemish Government, which is limited to 60% of income. Research funding can come from different local and federal government departments. Other sources are contract research from industry and international scientific initiatives such as the European Space Agency. Contract research from industry accounts for about half of the other sources. This source of income, particularly from industry in Flanders, is becoming more important; in 1992, 31.4% of the total contract research income originated from Flemish industry (IMEC Scientific Report 1992). While industry money has been responsible for the growth of the laboratories research activities, government money provided the starting point and continuing support – without which IMEC cannot succeed in reaching its objectives.

IMEC'S STRATEGY

IMEC is both an incubator and an agent of regional development. Its strategy is enacted along four major lines:

(1) Scientific research in IMEC runs 5–10 years ahead of industrial needs, preparing technological possibilities which can be directed towards existing industrial companies. By collaborating with industrial top-performers world-wide, IMEC aims to build up its reputation as a 'centre of excellence' in the field of micro-electronics – particularly in application specific integrated circuits (ASICs).

(2) Performing dedicated and flexible training in the field of very large scale integrated circuits (VSLI) chip design for educational as well as for industrial needs.

(3) Reinforcing the industrial research activities of Flanders. This is done in different ways:

- by creating spin-off companies
- by attracting foreign investment into the region by providing an investment package. To qualify for assistance, companies have to have R&D activity. This value-added development activity is seen as encouraging anchorage to the region.
- by transferring technology into industry (for example in sub-micron-process technology)
- by the supply of skilled personnel to local industry, as R&D staff leave at the end of their contracts.

(4) Performing strategic research in collaboration with universities.

The role of the laboratory is evolving as the emphasis and expectations from the government have changed. Whereas at the outset IMEC was primarily a research organisation undertaking long-term research, it now has responsibilities for helping SMEs in Flanders.

INCUBATION

Spin-off is part of IMEC's regional economic development missions. A firm has been formed from IMEC almost every year since 1986 (Table 12.1). With one exception, they have so far remained small. A barrier to more firms being formed is that some of IMEC's research results have no obvious immediate market.

UCB was the first IMEC spin-off. It is a small Belgian chemical company. It has invested money in equipment to produce an optical lithography process. IMEC had developed and patented a new method of etch which was more rapid than the standard and was entropic at a lower frequency. A second firm, UCB Electronics, was created to exploit this technique and is now valorising products world-wide.

In 1987, COBRAIN was formed. This company has a double IMEC input: seed capital funding and investment by the founders. It also had funding from the Flanders venture capital company (GIMV). By 1990, it had ten employees.

Table 12.1 Spin-offs from IMEC

Name	*year formed*	*employees*
UCB	1986	15 (1990)
Cobrain	1987	10 (1990)
European Development Centre (EDC)	1988	200 (1994)
Soltech	1989	8 (1992)
EASICS	1992	n/a
Destin	1992	n/a
Alphabit	1992	n/a

Source: IMEC 1993

The largest spin-off firm is European Development Centre (EDC). It was formed as a combination of spin-off and foreign investment. It was set up in 1988 and by 1993 employed over 200 people. It was formed to make engineering tools developed in IMEC. The company's core technology came out of a collaboration, under ESPRIT, with Philips in Eindhoven. When the research came to maturity, it needed industrial input to make the results into commercial products. An engineering company, EDC, was created with the American company SCS. The arrangement was that IMEC's contribution would be technology and IMEC would have the right of first refusal against SCS shares. IMEC is a symbolic shareholder, owning 5% of the shares.

SOLTECH was set up in 1989 to exploit full process crystalline photovoltaics cells developed at IMEC. In 1992 it employed eight people. IMEC is a shareholder and there are three industrial shareholders.

EASICS was formed in 1992 as a joint initiative of IMEC and KUL. EASICS focuses on two categories of activity: design projects, with the specification provided by the customer, and consultancy projects, using their experience of design methods, and of practical design and synthesis.

DESTIN was established in 1992 for the commercialisation of a new IMEC technology for the rapid measurements of ageing phenomena in electrical and electronic components and systems.

Some former IMEC researchers have joined spin-off organisations such as Cobrain, Soltech and UCB.

TRAINING

The most important formal training mechanism for which IMEC is responsible is INVOMEC. This programme provides training and education in design, providing software and training to use the software in design for schools, universities and industry. INVOMEC has contacts and networks with universities and schools for industrial engineers and 13 high schools in Belgium. These are all linked to the central computer at IMEC. The Multi-Project Chip Service (MPC), supported by INVOMEC, is a low-cost ASIC prototype fabrication service. Engineers work on test-case designs and prototypes through the MPC service. From 1986 the MPC fabrication was extended to all universities in Europe, research institutes and industry. Over 500 designs have been fabricated. Most have been for industry (214) and some for the Flemish Universities (143), Flemish Polytechnics (60) and IMEC (75) (IMEC Scientific Report 1992). INVOMEC is also active in the EC ESPRIT EUROCHIP programme providing training and software in design methodologies and trains some 300 VLSI engineers per year.

SMEs

INVOMEC's remit includes training for SMEs. The strategy to help SMEs goes beyond INVOMEC (see also van Tulder 1991). They can be given assistance with design and prototype development through consultancy and IMEC's servicing capability and test facilities. This offers a cheap way of processing chips for smaller firms. IMEC's strategy is to find common ground between it and small firms. By 1991, IMEC staff had visited 110 Flemish SMEs to discuss their problems and how IMEC's expertise can be incorporated into the firms. Working with IMEC also provides the opportunity for firms to be exposed to information about future technological trends and to meet foreign firms.

However, IMEC has identified a number of factors which influence whether technology transfer can take place. A key factor in determining whether SMEs and IMEC can work together is cost effectiveness from the point of both IMEC and the firm. There are other problems related to communication problems, psychological barriers and mismatches in the skill and technology bases of firms and that of IMEC:

- Problems in communication. SMEs are often started by one man with a brilliant idea, who is often not university trained. IMEC finds that it can be difficult to encourage such firms since they lack trained people with whom discussions could take place.
- Psychological barriers. They are generally present in firms when the R&D personnel have been trained to work on PCB surface-mounted devices and are unfamiliar with design methods for ICs.
- Some SMEs in Flanders, which are innovative on the system side (e.g. in speech processing and modems and telecommunications) and have the advantage of being technologically flexible, are reluctant to collaborate on R&D and miss out on the opportunity of sharing costs.
- The problem of scale. Integrated Circuit Computer Aided Design (ICCAD) tools are developed for ICCAD foundries, but these are not the smaller firms. A basic mismatch exists: IMEC, as an R&D institute,

operates at a different level of technological sophistication to most local firms.

IMEC believes that it cannot serve its first goal of improving the local industrial tissue without achieving its second goal, that of undertaking state-of-the-art research. Its view is that it is not realistic to set up an institute just for SMEs and to expect it to be supported by local industry. IMEC therefore achieves the first goal by creating a critical mass and creating inter-networks useful for SMEs. New breakthroughs with large companies are the drivers which open the way for followers. Therefore, success at the European level best serves the local level.

FOREIGN INVESTMENT

Another of IMEC's strategies is to attract R&D-intensive foreign industry into Flanders. It has already had a significant impact on the location of R&D activities of foreign companies. For example, Alcatel has moved the majority of its R&D into Flanders because of IMEC; and there are certain research departments of Philips which have remained in Belgium because of IMEC.

In order to encourage inward investment of the right kind, IMEC needs support from national and regional economic and investment policies. However, even in Flanders there are bureaucratic problems. It is IMEC's view that there are institutional rigidities within the local system. The difficulties arise with the complexity of the process of obtaining assistance from the public authorities. It might be supposed that under a locally responsive federal system, this would imply a more flexible attitude and more pro-active approach. In reality, Belgium is still in the process of moving from a national to a federal system. This means that the old bureaucracy has not disappeared while, at the same time, the local administration is in place. The consequence is that it is currently very difficult for firms to see which of the various authorities will resolve their problems.

SCK

SCK employs 560 people. Major changes in philosophy towards commercialisation of research developed in the late 1980s. For example, the physical separation of non-nuclear from nuclear activities occurred in 1991 with the formation of the Vlaamse Instelling voor Technologisch Onderzoek (VITO) on an adjacent site. VITO's primary commercial activities are industrial research on energy, environmental and biotechnologies and advanced materials. It was created as a Flemish rather than a national organisation.

Spin-off is part of current thinking at SCK. One firm, Identity, has been set up to exploit a number of ideas on fibre optics. The aim is to encourage the creation of others. It is recognised in SCK that, in order for this to be achieved, there has to be more flexibility and mobility on the part of personnel and this requires a change of culture. In order to encourage personnel to be more active in the exploitation of technology, SCK has adopted a position in which both the individual inventor and SCK profit from the intellectual property used to develop the firm. Although there have been some changes in philosophy and practice, the potential contribution of both organisations to local economic development is limited. Like the UK laboratories, SCK and VITO are geographically isolated from centres of industrial activity – they are located in a remote part of Belgium, near the border with The Netherlands.

To sum up, the evidence of a combination of IMEC and the growing Regional public sector infrastructure, which supports indigenous development and which aims to attract and retain higher, order activities of multi-national firms, suggests that Flanders is in the process of generating an innovative milieu or developmental scenario in which IMEC acts as a 'specialist supplier' of technology. This is in the form of direct technology transfer through research collaboration, recruitment by industry of ex-IMEC personnel, contributing to new firm formation, attracting skilled people into the region and the training it provides to local firms. The preceding discussion shows that, in a number of respects, the strategy has been successful – because of favourable conditions created by a raft of public intervention mechanisms. On the other hand, it is limited by local factors. Moreover, IMEC operates primarily on the international stage, which means that there is a considerable outflow of know-how and information rather than an inflow into new and/or indigenous firms. At the same time it is linked into international networks – which also act an antennae, for itself and firms in the region, detecting emerging trends in innovation.

France

This section focuses on the Laboratoire D'electronique de Technologie et Instrumentation (LETI) in Grenoble. It first sketches some of the characteristics of the national system of innovation in order to provide some historical and geographical context. This is followed by discussion of the laboratory's contributions to local economic development.

France is the fourth largest OECD economy, behind the United States, Japan and Germany. France is divided into 22 Regions (created in 1972), each of which encompass several departments (92 in total). Since World War II, the state has played a considerable role in shaping industrial development in France. Indeed, 'The French national system of innovation consists to a large extent of vertically structured and strongly compartmentalised *sectoral subsystems* often working for public markets and invariably involving an alliance between the State and public and/or private business enterprises belonging to the oligopolistic core of French industry' (Chesnais 1993, p.192). Chesnais argues that the French system has a number of serious weaknesses, 'expressed today in the endemic vulnerability of the French trade balance as well as the strong rigidity in the system in face of contemporary requirements for technological change' (p.193).

As part of measures to overcome the systemic barriers to innovation, France has adopted policies designed to foster economic development at the regional level: 'The directly elected Regional Council, and an Economic and Social Council in the Regions, serve chiefly as focal points for channelling government funds to different departments and, increasingly, to encourage inward investment in their area' (DTI 1991, p.4). Other agencies which operate to encourage exploitation of French national research at the regional level include: Agence nationale pour la valorisation de la recherche (ANVAR) formed in 1979 in order to find industrial partners for CNRS and university laboratories and Centres regionaux d'innovation et de transfert de technologies (CRITT) formed in 1982. The latter are joint venture organisations with public (mainly regional) and private financial participation and the job of enhancing regional innovation-enhancing networks between laboratories, firms and local government (Chesnais 1993). Moreover, the expansion of small- and medium-sized enterprises is now a major objective of Government policy (Innovation and Technological Development, ANVAR, August 1994).

However, the French innovation system is spatially polarised, which has meant that few areas currently have the capacity to develop innovative milieu. The country's scientific base is divided into Paris (being the headquarters of most government agencies) and the provinces (essentially the South and the South-East), to which successive decentralisation policies have dispersed activities (Beckouche 1991). Within these have developed important secondary centres, such as Grenoble.

Grenoble has a number of initial conditions which favoured the growth of an innovative milieu and one where government-funded research can be appropriated by local industry (see Hilpert and Ruffieux 1991). First, it has a long-established industrial base onto which newer cycles of investment have been grafted. The city developed in the early part of the twentieth century as an important industrial centre. Hydro-electric power enabled paper and cardboard manufacturers, metal-working and textile industries to become firmly established and later specialised in the manufacture of heavy electrical apparatus for hydro-electric power stations. Second, it has a remarkable concentration of education and training functions; three universities, the Institut National Polytechnique de Grenoble (INPG) plus grande ecoles and technical colleges (Dunford 1989). Third, the electronics industry in Grenoble has been primarily indigenous: two firms which existed in Grenoble prior to 1955 have been important with regard to the birth of new firms (Hilpert and Ruffieux 1991). Fourth, a series of government decisions on decentralisation of public sector institutions dating back to the 1950s reinforced and extended the city's scientific and engineering base. In 1956 CEA moved one of its five civil centres to Grenoble, which resulted in the creation of the Centre d'Etudes Nucleaires et de Grenoble (CENG). As a consequence of these accumulated factors, Grenoble has a concentration of highly skilled workers. By 1983, Grenoble had one per cent of the active French population, yet 14% of French graduates in electrical engineering qualified in the city and 10% of research jobs in the filiere electronique (Dunford 1989).

LETI

LETI is the main part of the 'advanced technology' programme of the Commissariat Energy Atomic (CEA), the French atomic energy authority. The main objective of the programme is the valorisation of the work which was initially done for nuclear energy. LETI was set up within CENG in 1967 and was formed as a result of a decision on what to do with the staff who had undertaken major developments in the nuclear reactor design when this had been completed. It was decided to employ these people on developments which would help (mainly French) industry. This has taken a variety of forms which are now examined:

INCUBATION

CEA funds LETI to encourage mobility and transfer of skills through a programme which supports the formation of new companies using LETI technology, and thereby generating employment. LETI has to demonstrate to CEA that this is a useful strategy, that is, there is a social advantage through employment generation, and that this is a more efficient way of exploiting the technology, compared to methods such as licensing technology. The spin-off arrangement provides security for entrepreneurs. When LETI employees leave to form companies, they remain employed by the CEA. The length of detachment is two years, with a renewal of up to five years. If there was not this safeguard, people would be less likely to leave – particularly in current times of high unemployment.

Between 1967 and 1992, nine spin-off firms were created – most since 1983. In 1993, another six were in the process of being formed. The fifteen firms are listed in Table 12.2. Some 127 staff have left LETI, of whom only 13 have been re-integrated. Etude et Fabrication de Circuits Integrés Speciaux (EFCIS) is a particularly notable start-up. In all 86 people have been attached to this company since its formation in 1972. Of these, 81 resigned to work full-time in the company and four returned.

All companies are created in market niches, usually special devices. Sometimes a new company develops under licence from LETI. In most cases the intellectual property remains in the ownership of LETI. For bigger companies and large markets there are normal technology transfer mechanisms. Some start-up companies operate inside LETI – their laboratories and offices are on site and they pay a fee to LETI for use of facilities. Close to the CEA site on which LETI is located is Astec, a special site developed by the CEA for new companies. On it are ten new spin-offs from LETI.

LETI recognises that the spin-off mechanism can cause tensions. There have been complaints from LETI scientists and engineers who found they cannot work as well as they could because they have to share their equipment with the new companies, which are under pressure to meet commercial deadlines. A solution to these problems has been for the industrial people to work a second shift after normal hours.

Table 12.2 LETI spin-off firms

Name of Company	*Number of People*	*Personnel Action*	*Year*
DAMELEC	1 person	1 resignation	1965
NOVELEC	1 person	1 resignation	1972
EFCIS	detachment of 86 people, significant departures in 1973, 1979, and 1980	4 reintegrations 81 resignations 1 deceased	1972–81
CRISMATEC	detachment of 7 people	4 reintegrations 3 resignations	1979
NEXTRAL	1 person	1 resignation	1983
NEF	1 person – firm creation	1 resignation	1985
SOFRADIR	detachment of 16 persons, 6 in progress	2 resignations 8 resignations	1986
ICAP	1 person, creation of firm	1 resignation	1987
C.S.O	2 persons	2 reintegrations	1987
In Progress			
ALPHA-PRIME	1 person: creation of the firm		1989
SERP	1 person: creation of the firm		1989
SILMAG	2 persons: creation of the firm		1991
ELDIM	1 person: creation of the firm		1992
SOITECH	4 persons: creation of the firm		1992
PIXEL INTERNATIONAL			1993

Source: LETI/DIR/G

OTHER MECHANISMS

Other links with the area are sustained by employees of LETI taking jobs in local electronics firms (more so in non-recessionary times), strong links with the local universities (reflecting common specialisations in micro-electronics and optoelectronics) and special programmes to support innovation in SMEs ('Diffusion Technologie').

CEA's mission is to help SMEs, even if it loses money in the process. Mechanisms include:

- giving technological advice
- identifying for SMEs when the solution to their problems can be found within CEA.
- helping SMEs identify the mechanisms to fund innovation
- directly part funding innovation. LETI can be involved in a project which is at a risky stage by sharing up to 50% of the expenses. The small company can get help for the other 50% from organisations including CRITT and ANVAR.
- innovation clubs.

Innovation clubs operate along similar lines to those operated in the UK. SMEs pay an entrance fee and have the right to come once or twice a year to a one-day meeting at LETI. They also have state-of-the-art reports on technological developments. Clubs are not a major activity for LETI. They tend not to be profitable and are more a service than a money earner.

In theory, the range of activities described are beneficial to local SMEs but, in practice, there are the same cultural and practical barriers to interaction found by IMEC. LETI finds that SMEs are very reluctant to innovate as innovation is risky and expensive, that a big organisation can be intimidating and that the high-level engineers who work in a big research laboratory do not understand the environment of small firms. One of the problems is that the research being done in LETI now will have industrial applications in five to ten years time, whereas industry is interested in production and needs product innovation rather than innovation in a more general sense.

This study suggests that LETI's enterprise strategies have a role to play in contributing an innovative milieu by the use of the skills of its personnel. However, as in Flanders, there are barriers to success – particularly the match of technological sophistication and the practical needs of firms.

Conclusions

This study has illustrated that, whereas the UK has been very slow to develop regional development strategies using the resources of national laboratories, France and Belgium are far ahead in recognising the possibilities of using institutions within a devolved spatial scale of economic development. So far the only one of the UK sample which is becoming involved in a regional strategy is NEL, and this is in an area with no historical basis for optimism for the development of an innovative milieu.

The range of methods by which technology transfer is facilitated in the Flanders and Grenoble examples indicate that far more complex processes are involved in the way institutions form part of the fabric of regions through the spatial integration of skills than has so far been recognised in the literature. In the examples of Belgium and France discussed above, what has actualised local systems of production (Alford and Garnsey 1994) has been state intervention, reinforcing trends established by earlier regulatory measures. The Grenoble region has had a history of industrial development underpinned by the state, while in Flanders, since the 1960s, policies have been orientated to achieving economic development based on technology transfer within regional specialisations. State intervention has, therefore, created favourable conditions which may now constitute innovative milieux, or possibly a series of sub-systems or sub-milieux, based on particular sectors. However, the issue of scale remains problematic. In this and in other studies it is not clear as to what extent the system of local technical linkage extends to include adjacent areas (that is, what are the boundaries of the milieu?) or to what extent the milieu is sustained by labour advantages of quality and skills (Keeble 1976) reinforced by actions of individual institutions.

The Flanders strategy of generating regional economic development through the means of the multi-functional organisation IMEC has succeeded in generating a match of interests between itself, educational institutions and some firms in Flanders. It was created out of a political vision which established a partnership between technological achievement and industrial application, and acts as an incubator and a practical resource for indigenous and foreign firms. However, IMEC suffers from 'institutional rigidities' within Flanders which interfere with the objectives of this imaginatively conceived institution, and from the lack of match of interests with local small firms, a problem recognised in numerous other studies.

While studies of innovative milieu are useful in identifying indicators of characteristics of certain types of regions, it may be time to concentrate on other forms of regional development not based on high-tech activities – which form only a small proportion of employment in most countries. For example, of the total 1991 UK work-force of nearly 24 million, less than 5% were employed in high-tech activities (Jordan 1995). High-tech activities are themselves only part of production system or *filiere*, 'a connecting filament among technologically related activities, (Truel 1980 quoted in Storper and Walker 1989), and the 'value-added' of location for even high-tech firms may not be a significant factor in industrial competitiveness.

The author would like to thank Dr G.P. Sweeney and Professor Mike Atkinson for comments on earlier versions of this chapter. The chapter appears in 'Entrepreneurship and Regional Development' No.1 Volume 8 1996, pp.1–17.

References

Alford, H. and Garnsey, E. (1994) 'Flexibility and specialisation in supplier relations among new technology based firms: An exploratory study in the Cambridge Area.' *Research Papers in Management Studies 1993–4 No. 13*. Judge Institute of Management Studies, University of Cambridge.

Beckouche, P. (1991) 'French high tech and space: a double cleavage.' In G. Benko and M. Dunford. *Industrial Change and Regional Development: the Transformation of New Industrial Spaces*. London: Belhaven Press.

Bianchi, P. and Giordani, M.G. (1993) 'Innovation policy at the local and national levels: the case of Emilia-Romagna.' *European Planning Studies 1*, 1, 23–42.

Camagni, R. (1993) 'Inter-firm industrial networks: the costs and benefits of co-operative behaviour.' *Journal of Industry Studies 1*, 1, 1–15.

Cheshire, P.C., D'Arcy, E. and Giussani, B. (1992) *Local, Regional and National Government in Britain. A Dreadful Warning*. Paper presented at Thirty-Second European Congress, Regional Science Association. Louvain-La-Neuve, Belgium, 25–28 August 1992.

Chesnais, F. (1993) 'The French National System of Innovation.' In R. Nelson (ed) *National Innovation Systems: A Comparative Analysis*. New York and Oxford: Oxford University Press.

Cohen, W.M. and Levinthal, D.A. (1990) 'Absorptive capacity: a new perspective on learning and innovation.' *Administrative Science Quarterly 35*, 128–152.

Cooke, P. (1993) *The New Wave of Regional Innovation Networks: Analysis, Characteristics and Strategy.' Regional Industrial Research, Department of City and Regional Planning, University of Wales, Cardiff, February 1993.*

Cooper, A.C. (1971) 'Spin-offs and technical entrepreneurship.' *IEEE Transactions on Engineering Management*, EM-18, 1, 2–6.

Department of Trade and Industry (1988) *Support for Club Projects Evaluation Report*. London: DTI.

Department of Trade and Industry (1991) *Country Profile: France*. London: DTI.

Dunford, M. (1989) 'Technopoles, politics and markets: the development of electronics in Grenoble and Silicon Glen.' In M. Sharp and P. Holmes (eds) *Strategies for New Technologies: Case Studies from Britain and France*. New York and London: Philip Allan.

Ergas, H. (1993) *Europe's Policy for High Technology. Has Anything Been Learnt?* Paris: OECD.

Ettlinger, N. (1994) 'The localisation of development in comparative perspective.' *Economic Geography 70*, 144–166.

Garnsey, E.W., Galloway, S.C. and Mathisen, S.H. (1994) 'Flexibility and specialisation in question: birth, growth and death rates of Cambridge new technology-based firms 1988–92.' *Entrepreneurship and Regional Development 6*, 81–107.

Heim, C.E. (1988) 'Government research establishments, state capacity, and distribution of industry policy in Britain.' *Regional Studies 22*, 375–386.

Hilpert, U. and Ruffieux, B. (1991) 'Innovation, politics and regional development: technology parks and regional participation in high tech in France and West Germany.' In U. Hilpert (ed) *Regional Innovation and Decentralisation: High Tech Industry and Government Policy*. London: Routledge.

Huggins, R. (1995) *Competitiveness and the Global Region: The Role of Networking*. Paper presented at Regional Studies Association Conference on Regional futures: Past and Present, East and West, Gothenburg, 6–9 May.

IMEC Scientific Report (1992) IMEC, Kapeldreef 75, Leuven, Belgium.

Jordan, S. (1995) *Industrial Change, Labour Mobility and Regional Economic Development: The Case of High Technology Industry in Oxfordshire.* Honours School of Geography Geographical Dissertation, January 1995.

Keeble, D. (1976) *Industrial Location and Planning in the United Kingdom.* London: Methuen.

Keeble, D. (1990) 'Small firms, new firms and uneven regional development in the United Kingdom.' *Area 22.3*, 234–245.

Lawton Smith, H. (1990) 'Innovation and technical linkages: the case of advanced technology industry in Oxfordshire.' *Area 22*, 125–135.

Lawton Smith, H. (1991) 'The role of incubators in local industrial development: the cryogenics industry in Oxfordshire.' *Journal of Entrepreneurship and Regional Development 3*, 2, 175–194.

Lawton Smith, H. (1994) *A Report on The Contribution of National Laboratories to the European Scientific Labour Market.* School of Geography, Mansfield Road, Oxford.

Lawton Smith, H. (1995) 'The contribution of national laboratories to the European scientific labour market.' *Industry and Higher Education 9*, 3, 176–185.

Lawton Smith, H. (forthcoming 1997) 'Regulatory change and skill transfer: the case of national laboratories in the UK, France and Belgium.' *Regional Studies 31*,1.

Longhi, C. (1995) *A Note on the Sophia-Antipolis Experiment: Development and Crisis.' Paper presented at the European seminar, Cambridge, 2–3 March, 1995.*

Mommen, A. (1994) *The Belgian Economy in the Twentieth Century.* London: Routledge.

Moulaert, F. and Willekens, F. (1987) 'Decentralisation industrial policy in Belgium: towards a new economic feudalism.' In H. Muegger, W. Stohr, P. Hisp and B. Stickey (eds) *International Economic Restructuring and the Regional Community.* Aldershot: Avebury.

OECD (1989) *The Changing Role of Government Research Laboratories.* Paris: OECD.

Palfreyman, D. (1989) 'The Warwick way: a case study of entrepreneurship within a university context.' *Entrepreneurship and Regional Development 1*, 207–219.

Saxenian, A. (1991) 'The origins and dynamics of production networks in Silicon Valley.' *Research Policy 20*, 423–437.

Sayer, A. and Walker, R. (1992) *The New Social Economy: Reworking the Division of Labour.* Oxford: Blackwell.

Segal Q.W. (1988) *Universities, Enterprise and Local Economic Development: An Exploration of Links.* Report for the MSC, HMSO, London.

Storper, M. and Walker, R. (1989) *The Capitalist Imperative.* Oxford: Blackwell.

Storper, M. (1993) 'Regional "worlds" of production: learning and innovation in the technology districts of France, Italy and the USA.' *Regional Studies 27*, 5, 433–456.

Turok, I. (1993) 'Inward investment and local linkages: how deeply embedded is "Silicon Glen"?' *Regional Studies 27*, 5, 401–418.

van Tulder, R. (1991) 'Small industrialised countries and the global innovation race: The role of the state in the Netherlands, Belgium and Switzerland.' In U.Hilpert (ed) *State Policies and Techno-Industrial Innovation.* London: Routledge.

PART V

Conclusions

CHAPTER 13

Summary and Conclusions

James Simmie

The research presented in this volume has addressed three broad concerns with respect to the relationships between innovation and regions. These revolve around such questions as: what are the characteristics which result in core metropolitan regions being the locations for such high proportions of all innovations? And, what strategies could be adopted in peripheral regions to overcome the disadvantages of such regions in the race for competitive economic change? Finally, the issue of how technological knowledge may be transferred and diffused as widely as possible is explored.

In Part I the author has reported that innovation in core metropolitan areas appears to be a more chaotic activity than some of the available theories suggest. There is not much consistency in the types of innovation discovered, nor is there much evidence of either systematic networking or high-level linkages contributing to award-winning innovations. The firms interviewed so far do not regard government regulations as contributing to innovation. Thus, theories that include networking, local industrial organisation or regulatory regimes do not appear to fit the Hertfordshire case.

A fluid system is emerging from the research results. In this context, numerous, seemingly unrelated, innovations have bubbled up from a complex knowledge-base which is embodied in the highly-educated, professional work-force that has chosen to live in and around Hertfordshire. The nearby availability of venture and long-term capital emanating from sources based in London is also a major contributory factor to innovations in the area. The successful innovating firms can, and do, compete in global markets from their base in Hertfordshire. In doing so they are in competition with other companies, mainly in North America and other world-wide locations.

Pooran Wynarczyk, Alfred Thwaites and Peter Wynarczyk suggest that significant innovations are more likely to be introduced into the South-East region than elsewhere in the UK. Their research also showed that retained profits and exports grow more strongly in firms located in that region than in those located in other regions of the UK.

The owners of innovative SMEs in the South-East were more willing and able to issue relatively large numbers of additional shares, which, in aggregate, were seen to double in the four years after innovation. At the same time, the owners were willing to let their proportion of equity holding reduce, with some giving up overall control. Such firms were usually more successful in export markets. This is a significant indicator of their relative international competitiveness.

Some of this international competitiveness was built on the greater willingness and ability of South-East SMEs to appoint technical and scientific directors. These professional directors were closely associated with exports and profitability growth, in contrast to the family run companies in other regions.

Jeanine Cohen analysed the development of R&D employment in the Paris region. This has been marked by the development of collaborative industrial policies between state agencies and international-scale companies. As a result, in the 1960s, with the taking off of R&D jobs; in the 1970s, with the constitution of big international-scale companies; and in the 1980s, with the actions to help create SMEs – especially innovative ones – French industrial policy seemed to be more and more committed to R&D. High-technology was the 'locomotive' of its economy.

But, since 1990, unemployment problems have returned. Rapid de-industrialisation, particularly of the centre of Paris, has started to slow the growth of related producer services in the city. Nevertheless, as with London, high-technology activities continue to survive, and even flourish, in the outer metropolitan areas such as the Saclay Plateau south of Paris and the western arc around London.

Heidi Wiig and Michelle Wood present research results on the characteristics of the local innovation system of Møre and Romsdal – a core region of central Norway. They argue that many firms there undertake innovation in products and processes. There is a strong regional economic environment and a specific type of innovation system.

Nevertheless, as with both London and Paris, there is little evidence to suggest interaction between firms for innovation. The presence of related firms is seen as unimportant to firms' activities and firms do not see other institutions as valuable sources of information, expertise and support for their innovation activities.

Taken together, all these findings cast some doubt on the importance and significance of local supply networks as a major contributory factor to innovation in core metropolitan areas. Firms in London, Paris and Møre and Romsdal, in different studies, have all contradicted some of the received academic wisdom on the importance, or even existence, of network contributions to innovation. This requires some further work and explanation.

Descriptively, at least, it appears to be the case that innovative firms in core regions often produce innovations in-house and without outside connections. This is facilitated, on the one hand, by their ability to employ the high-quality professional labour required to develop applied scientific knowledge. On the other hand, concerns with secrecy and commercial advantages also lead firms to protect their activities by not sharing them with anyone outside the firm. Where a lot of innovation is being conducted in innovative core regions, the need to search outside individual firms for inventions and knowledge is, perhaps, less than in firms and regions where innovation is more scarce.

Conversely, the relative export success of core metropolitan regions is at least correlated with links to foreign customers. These are often sector or niche specific. Such regions may therefore be characterised by a series of separate and distinct networks which connect them to export customers in different sectors and regions of the global economy. In this sense, the companies located in such regions are adept at operating both locally and globally at the global/local interface.

Heidi Wiig and Michelle Wood conclude from their study that there is a major problem in constructing a definition of an innovation system which is applicable to the whole range of different localities and regions. The research reported in Part I of this book suggests that individual firm strategies and networks actually work against the formation of visibly-integrated regional innovation systems.

In Part II, Robert Huggins reiterates the position that an important element of recent theory in the field of regional dynamics is based on network concepts – in terms of information, knowledge and innovation networks favouring the competitive advantage of regions. A capacity to network, which ties a region to relevant external partners, may therefore become a stronger determinant for development than many other previously important internal factors.

Despite this important distinction between internal and external networking activities, his report of attempts to develop networks in South Wales shows that they are often focused more on overcoming the problems of geographic peripherality and internal supply issues. The study of the main types of services required by firms showed that they were partner-finding for R&D, information on technology developments, partner-finding to obtain finance and advice on product design and development. The primary users of the South Wales network were argued to be SMEs who are already seeking partners on a regional or national level. Some, however, might also be seeking to supply collaborators on a European or world-wide basis.

The most serious constraint on innovation performance, particularly among smaller firms, was funding and the cost of finance. The importance of cost factors in constraining the introduction of new technologies reinforces the view that SMEs, especially in regions outside the South-East, are severely disadvantaged by the cost of innovating and the difficulties of raising funds for innovation.

Andy Pratt raises some important questions about the perceived utility and neutrality of networks in general. He contends that the power/institution argument counters the idea of institution as context or as a meso-level intermediary; it also resists the representation of networks as neutral or cloaks of power. It does not presume that agents (which may be individuals, firms or regions) come pre-formed or ready made. Its central concern is with the ways in which agents are constructed and translated in practice. It is also concerned with what the effects of such translations are in terms of power.

This raises two critical issues for peripheral regions in particular. The first concerns the location of power within such regions. Existing relations are often those that have generated uneven economic development in the first place. The development of local supply networks needs to beware of re-inventing these relationships with respect to new industries.

Second, building a network, a science park, or a better mousetrap, may be a necessary condition for local economic regeneration but it is not sufficient unless others can be convinced of its value and enrolled into its promotion and use. Furthermore, non-specific activities do not necessarily produce anything that can be sold commercially and competitively in global markets. As such, they do not count as innovations in the strict definition of that concept.

Sang-Chul Park and Rolf Sternberg have analysed the massive Japanese technopolis policy to develop the more peripheral regional economies in that country on the basis of knowledge-intensive industries. They show that there is no substitute for long-term strategic planning in terms of science and technology, industrial and regional policies backed up with large-scale public funding and private investment to develop peripheral economies. Even then, there is no guarantee of immediate success or shaking off the dominance of core metropolitan regions.

Sang-Chul Park outlines the technopolis programme. Its intention is that the weakness of basic research, compared to Western countries, should be overcome by corporatist collaborations between government, industry and the research community. The technopolis plan was introduced by MITI in 1980. Some 26 technopolises have been designated.

After over a decade of such major efforts, in 1991, 20 of the entire 26 technopolises have only reached their first step in completing the infrastructure and facilities for their planned R&D activities. As yet, he argues, it is too early to assess the results of the technopolis plan. They will start to come out slowly at the end of the 1990s.

Rolf Sternberg has evaluated the results of the Kyushu technopolis so far. He concludes that it deserves the predicate 'Silicon Island' insofar as its relative number of high-tech employees does indeed far exceed the national average. However, this technology intensity is almost exclusively based on one industrial branch, the semi-conductor industry. On the other hand, the term 'Silicon Colony' also applies because the high-tech businesses on Kyushu depend heavily on the core metropolitan region of Tokyo-Osaka-Nagoya. He is pessimistic about the potential of the technopolis programme to overcome the spatial disparities of this core region and peripheral areas. In addition, the relative success of the individual technopolis is shown to decrease with distance from the central core.

The main comparative advantage enjoyed by Kyushu is the supply of relatively low-cost and well-trained labour. This has not been sufficient, however, to create an endogenous development dynamism with strong intra-regional linkages and an innovative milieu. Furthermore, the role of central government in terms of finance, organisation and political support is still critical for the potential future success of all the Japanese technopolises.

The conclusions of Parts I and II show that, while there are a number of necessary conditions for innovation-driven economic growth – such as the availability of knowledge and information, public and private commitment to incorporating them in strategies, adequate funding and possibly some kinds of networks – none of these are sufficient conditions to guarantee success. The research in Part III, therefore, concentrates on the question of the successful use of knowledge and its transfer and diffusion as a condition of economic growth.

Nicos Komninos argues that spatial characteristics are becoming less significant with respect to innovation. He concludes that the spread of new effective tools for technology transfer, based on networks, institutions and services, questions the established character of technopolitan development. Many of them are not space dependent. In theory, networks and information channels do not presuppose the spatial proximity of the participating members and open their linking capacity over large geographical scales.

Keith Tanner and David Gibbs report a study of UK local authority attempts to facilitate such non-spatial technology transfer using information and communication technologies. They showed that around a third of local authorities already had an ICT policy, while slightly more than a further third were also considering the development of such a policy. In many cases these policies were partly concerned with overcoming perceived peripherality disadvantages.

They went on to point out that it is still rare to see any analysis of exactly how such ICT strategies will have a directly beneficial outcome for the local economy. Strategies have developed in a piecemeal fashion and often without a clear idea of the economic development benefits they are supposed to achieve.

The rather diffused objectives of those local authority initiatives in operation were to: provide access to information, provide training – especially for unemployed people and disadvantaged groups – and to encourage business development, especially in the form of encouraging small-firm networking and networked workspace. There was little focus on innovation in all these efforts.

Again, adopting a supply-side approach and creating better infrastructure provision may be a necessary, but not sufficient, cause of increased economic activity. Policy makers cannot assume that the provision of ICTs is a sufficient condition for economic advance within a city or region. In relation to SME networking, for example, the network between firms must exist before it can be mediated through ICTs and policy cannot create networks where none existed before, simply through providing an ICT infrastructure.

The assumption that networking will occur and the implicit assumption that some form of local production complex or industrial district will result appears to underlie much local authority ICT policy directed to SMEs. The evidence is, however, that ICT usage is more likely to promote the processes of economic globalisation rather than supporting the re-emergence of neo-Marshallian industrial districts. This might not be such a bad result if it connected localities with export customers in global markets.

The research reported in Parts I and II of this volume cast some doubt on the general applicability of networking and industrial district theories. Keith Tanner and David Gibbs replicate the findings in Hertfordshire that small firm participation in ICT networks may be motivated by customer pressures rather than the desire or ability to interlink with other local small firms, which are more likely to be seen as competitors than collaborators. Further research questions remain as to whether such fragmented local competition is a peculiarly metropolitan, or even, UK phenomenon. The evidence from London, Manchester, Paris and Møre and Romsdal reported here suggests that it may be a more general metropolitan characteristic than many networking and new industrial district theories have proposed so far.

Keith Tanner and David Gibbs also report that the commonly missing ingredient in most UK initiatives is significant public funding from central government. Both the French and Japanese research reported in this volume suggest that significant long-term central government strategic funding is a crucial element in the development and redevelopment of knowledge-based economies at any spatial level. Local authorities, TECs and even the European Union have not, so far, been able to substitute for inadequate central government funding for innovation and its prerequisites in the UK.

Finally, Helen Lawton Smith presented evidence from the UK, France and Belgium showing the importance of people movement with respect to technology transfer. Her evidence supports the contention that technological solutions on their own are not a sufficient cause of technology transfer. It is important to draw a distinction between information and knowledge. Information can be collected and dispersed in many ways, including the use of information technologies. Despite this, it requires the knowledge that resides in peoples' heads to recognise the significance of information and use it in successful innovation.

Helen Lawton Smith examines the roles of national laboratories in transferring technologies by means of encouraging knowledgeable individuals to move out of them and to generate local technological spin-offs. She shows that the UK has been very slow to develop regional development strategies using the resources of national laboratories. In contrast, France and Belgium are far ahead in recognising the possibilities of using public institutions to spin-off local economic development.

In the two latter countries, what has actualised local systems of production has been state intervention – re-inforcing trends established by earlier regulatory measures. State intervention has created favourable conditions which may now constitute innovative milieux, or possibly a series of sub-systems or sub-milieux, based on particular sectors. The Flanders strategy, for example, of generating regional economic development through the means of the multi-functional organisation IMEC, has succeeded in generating a match of interests between itself, educational institutions and some firms in Flanders.

Helen Lawton Smith also draws attention to the fact that, while studies of innovative milieu are useful in identifying indicators of characteristics of certain types of regions, it may be time to concentrate on other forms of regional development not based on high-tech activities – which form only a small proportion of employment in most countries. For example, of the total 1991 UK work-force of nearly 24 million, less than 5% were employed in high-tech activities.

Taken together, the contributions to this volume have identified a number of key issues with respect to innovation and the regional question. The first of these is the need to understand the special roles of core metropolitan regions with respect to national innovation. The evidence presented above shows that, on the one hand, they tend to be the most innovative geographic concentrations. On the other hand, they show few, if any, of the characteristics hypothesised in network and new industrial district theories. Far from being the locations of collaborative and socially-embedded supply-side networks, they seem to contain secretive and competitive firms whose networks are much more likely to be with demand-side export customers. The ability to produce local actions which make them competitive in global markets is an important indication of their innovative success.

The second key issue is that the more peripheral Western regions examined here have all shown interest in overcoming their peripherality by developing local supply-side networks and information technologies. In doing so they may be misunderstanding what makes more central regions successful, in terms of their abilities to innovate continuously. One of the key features of the latter is the concentration of highly-qualified professional workers. It is mostly such people who learn and innovate. Regions, networks, information technologies and new industrial districts are inanimate objects which, by themselves, can do neither of these two things.

Even highly-qualified professional workers are limited in their abilities to innovate without adequate public and private funding. The Japanese technopolis programme aims to bring all these ingredients together in many of the country's more peripheral regions. Even so, it has been shown that massive and very long-term efforts are still required to overcome the innovative advantages enjoyed by the central core megalopolis. This indicates that peripheral regional disadvantage cannot be overcome cheaply. A few supply-side networks or improved information technologies will not be enough, on their own, to bring lagging regions up to the economic performance of their national core areas.

The third and final major issue concerns the importance of technology transfer and diffusion and how best to achieve it. This brings us back to the distinction between information and knowledge. Much of the evidence shows that policy makers often proceed on the basis that information transfer by technological means is sufficient to bring about significant technology transfer. This is seldom the case. Knowledge contained in highly-qualified professional workers' heads is an essential ingredient of real technology transfer. Thus attempts to generate spin-offs by encouraging staff to move out of government-funded research establishments may be a better example of real technology transfer.

Developing local concentrations of high-level professional workers and spinning them off into local firms looks like a more effective way to diffuse innovative capacities in local economies than developing hardware. Even so, a note of caution is in order in so far as neither a majority of national economic development nor a large part of the work-force can be based solely on high-technology activities. At the moment there are simply not enough to go round.

The Contributors

Jeanine Cohen, Stratégies Territoriales et Dynamique des Espaces, CNRS-Universite Paris I.

David Gibbs, School of Geography and Earth Resources, University of Hull.

Robert Huggins, Centre for Advanced Studies in the Social Sciences, University of Wales, Cardiff.

Nic Komninos, Department of Urban and Regional Planning and Development, School of Engineering, Aristotle University, Thessaloniki.

Helen Lawton Smith, Centre for Local Economic Development, Coventry University.

Sang-Chul Park, Centre for East and Southeast Asian Studies, Gothenburg University.

Andy Pratt, Department of Geography, London School of Economics and Political Science.

James Simmie, Bartlett School of Planning, University College London.

Keith Tanner, External Funding Manager, Burnley College.

Rolf Sternberg, Department of Economic and Social Geography, University of Cologne.

Alfred Thwaites, Centre for Urban and Regional Development Studies, University of Newcastle.

Heidi Wiig, STEP Group (Studies in technology, innovation and economic policy), Oslo.

Michelle Wood, Centre for Urban and Regional Development Studies, University of Newcastle.

Peter Wynarczyk, Department opf Economics University of Northumbria.

Pooran Wynarczyk, Centre for Urban and Regional Development Studies, University of Newcastle.

Subject Index

Name Index

For Product Safety Concerns and Information please contact our EU representative GPSR@taylorandfrancis.com Taylor & Francis Verlag GmbH, Kaufingerstraße 24, 80331 München, Germany